*More praise for*

# THE MIND OF A BEE

"*The Mind of a Bee* is a fascinating book that I hope will be read and understood by as broad an audience as possible, so that the important conclusions within may be shared more widely."

—AMANDA WILLIAMS, Buzz about Bees

"This is an amazing book. I give it my highest recommendation."

—DAVID GASCOIGNE, *Travels with Birds*

"This is an outstanding book that provides a comprehensive overview of honeybee cognition. It provides a clear introduction to the field for amateur bee lovers as well as a nuanced and up-to-date summary for professionals. By looking at the world through the lens of a bee, readers will develop tools to better understand the distinct and vivid experiences of tiny invertebrates that are too frequently ignored."

—ELIZABETH A. TIBBETTS, *Current Biology*

"[A] devoted, accessible analysis."

—ANDREW ROBINSON, *Nature*

"Quite simply a magnificent book. No one knows the mind of a bee better than Lars Chittka. A satisfying blend of sound science and spellbinding storytelling. I was mesmerized from the start."

—GEORGE MCGAVIN, zoologist, author, and broadcaster

"Lars Chittka's *The Mind of a Bee* is a mind-blowing presentation of scientific evidence and insight showing beyond any reasonable doubt that bees have awareness, memories, basic emotions, intelligence, and personalities—and that what we are doing to them and their world has not just practical but moral implications."

—CARL SAFINA, author of *Beyond Words* and *Becoming Wild*

"The time that insects were seen as little machines, incapable of complex thought, emotions, and learning, is far behind us. We can wish for no better guide than Lars Chittka for an accessible introduction to the wonders of bee intelligence."

—FRANS DE WAAL, author of *Are We Smart Enough to Know How Smart Animals Are?*

"Thanks to Lars Chittka's captivating account of bees' thinking and feeling, I now look with fresh eyes at these small animals. They plan ahead, feel pain, and express their very own personalities. The ingenious experimental evidence Chittka offers in support of these and many other points is as convincing as it is fascinating."

—BARBARA J. KING, author of *Animals' Best Friends: Putting Compassion to Work for Animals in Captivity and in the Wild*

"Lars Chittka is an ideal guide to the rich sensory world of bees and to their surprisingly sophisticated powers of cognition. Beautifully illustrated and filled with insights from decades of research, *The Mind of a Bee* combines scholarship and storytelling in nothing less than a tour de force. Highly recommended for any serious bee enthusiast!"

—THOR HANSON, author of *Buzz: The Nature and Necessity of Bees*

"Intensely detailed, meticulously researched, and always illuminating, *The Mind of a Bee* is as enjoyable as it is intellectually stimulating. This book takes a fascinating deep dive into bees' lives and minds, raising critical new questions for us as a species."

—HELEN JUKES, author of *A Honeybee Heart Has Five Openings*

# THE MIND

OF A

# BEE

# THE MIND

OF A

# BEE

LARS CHITTKA

Princeton University Press
Princeton and Oxford

Published by Princeton University Press
41 William Street, Princeton, New Jersey 08540
99 Banbury Road, Oxford OX2 6JX

press.princeton.edu


First paperback printing, 2023
Paperback ISBN 978-0-691-25389-3
Cloth ISBN 978-0-691-18047-2
ISBN (e-book) 978-0-691-23624-7

British Library Cataloging-in-Publication Data is available

Editorial: Alison Kalett and Hallie Schaeffer
Production Editorial: Jenny Wolkowicki
Text design: Wanda España
Jacket/Cover design: Lauren Michelle Smith
Production: Jacqueline Poirier
Publicity: Sara Henning-Stout and Kate Farquhar-Thomson
Copyeditor: Annie Gottlieb

Jacket/Cover image: Orchid bee, *Euglossa sp. Euglossini*. © Andreas Kay

This book has been composed in Warnock Pro and Gotham

Printed in the United States of America

For Marlene Sámano Chong

# CONTENTS

# 1 Introduction

Let us suppose that an inhabitant of Venus or Mars were to contemplate us from the height of a mountain, and watch the little black specks that we form in space, as we come and go in the streets and squares of our towns. . . . All he could do, like ourselves when we gaze at the hive, would be to take note of some facts that seem very surprising; and from these facts to deduce conclusions probably no less erroneous, no less uncertain, than those that we choose to form concerning the bee. . . .

"Whither do they tend, and what is it they do?" he would ask, after years and centuries of patient watching. "What is the aim of their life, or its pivot? . . . I can see nothing that governs their actions. The little things that one day they appear to collect and build up, the next they destroy and scatter. They come and they go, they meet and disperse, but one knows not what it is they seek.

**—Maurice Maeterlinck, 1901**

Understanding the minds of alien life-forms is not easy, but if you relish the challenge, you don't have to travel to outer space to find it. Alien minds are right here, all around you. You won't necessarily find them in large-brained mammals—

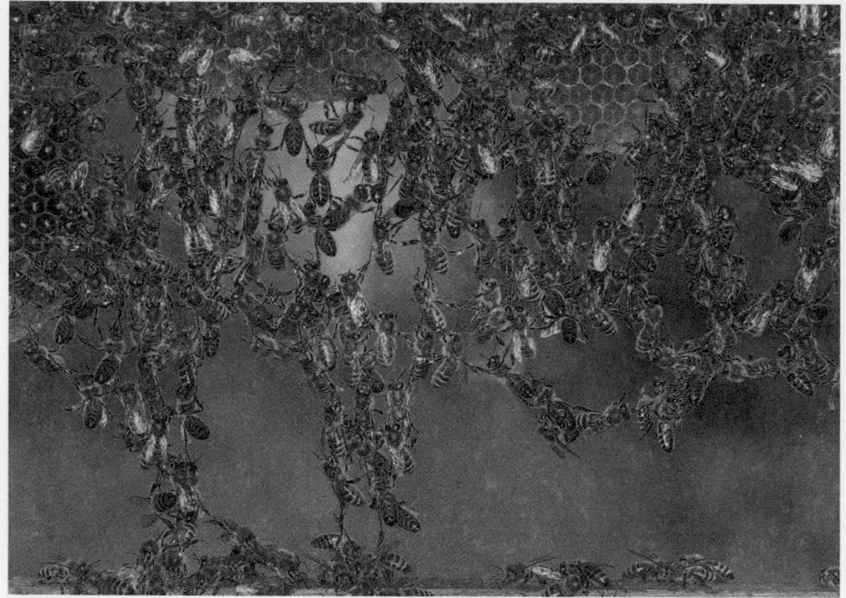

**Figure 1.1. The strangeness of the bee's world.** Many aspects of a bee's life, or its communities, have no parallel in the human realm. Unique forms of sensory perception, instinctual behavior, cognition, and social interaction can give rise to structures such as the mathematically optimal honeycomb, matchless in the animal kingdom in terms of its regularity and functionality.

whose psychology is sometimes studied for the sole purpose of finding human-ness in slightly modified form. With insects such as bees, there is no such temptation: neither the societies of bees nor their individual psychology are remotely like those of humans (figure 1.1). Indeed, their perceptual world is so distinct from ours, governed by completely different sense organs, and their lives are ruled by such different priorities, that they might be accurately regarded as aliens from inner space.

Insect societies may look to us like smoothly oiled machines in which the individual plays the part of a mindless cog, but a superficial alien observer might come to the same conclusion about a human society. Over the course of this book, it will be my goal to convince you that each individual bee has a mind—that it has an awareness of the world around it and of its own knowledge, including autobiographical memories; an appreciation of the outcomes of its own actions; and the capacity for basic emotions and intelligence—key ingredients of a mind. And these minds are supported by beautifully elaborate brains. As we will see, insect brains are anything but simple. Compared to a human brain with its 86 billion nerve cells, a bee's brain may have only about a million. But each one of these cells has a finely branched structure that in complexity may resemble a full-grown oak tree. Each nerve cell can make connections with 10,000 other ones—hence there may be more than a billion such connection points in a bee brain—and each of these connections is at least potentially plastic, alterable by individual experience. These elegantly miniaturized brains are much more than input-output devices; they are biological prediction machines, exploring possibilities. And they are spontaneously active in the absence of any stimulation, even during the night.

## What It's Like to Be a Bee

To explore what might be inside the mind of a bee, it is helpful to take a first-person bee perspective, and consider which aspects of the world would matter to you, and how. I invite you to picture what it's like to be a bee. To start, imagine you have an exoskeleton—like a knight's armor. However, there isn't any skin underneath: your muscles are directly attached to the armor. You're all hard shell, soft core. You also have an inbuilt chemical weapon, designed as an injection needle that can kill any animal your size and be extremely painful to animals a thousand times your size—but using it may be the last thing you do, since it can kill you, too. Now imagine what the world looks like from inside the cockpit of a bee.

You have 300° vision, and your eyes process information faster than any human's. All your nutrition comes from flowers, each of which provides only a tiny meal, so you often have to travel many miles to and between flowers—and you're up against thousands of competitors to harvest the goodies. The range of colors you can see is broader than a human's and includes ultraviolet light, as well as sensitivity for the direction in which light waves oscillate. You have sensory superpowers, such as a magnetic compass. You have protrusions on your head, as long as an arm, which can taste, smell, hear, and sense electric fields (figure 1.2). And you can fly. Given all this, what's in your *mind*?

## The Challenges of Being a Forager in the Wild

What is in an animal's (including a human's) mind is a mixture of information from its evolutionary history; information passing through the sensory filters it has acquired during evolution; information it has memorized from its experience; and things it might imagine, or anticipate. To explore the possible contents of a mind, it helps to think about what matters to the animal in question—what's important in that animal's daily life. For example, one thing you can be fairly sure is *not* on a honey bee worker's mind is sex: worker bees are typically sterile, and female reproduction is ceded to the queen. On the other hand, flowers are likely to have

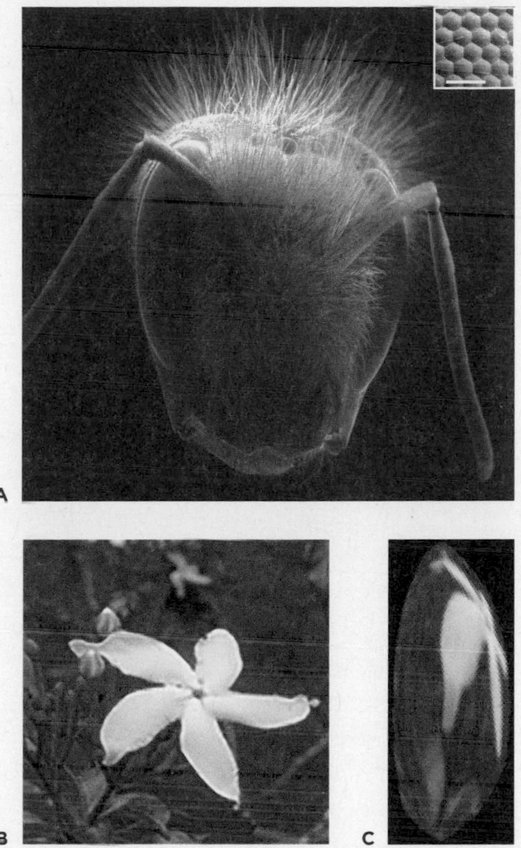


**Figure 1.2. Portrait of a bumble bee, and how it may see a flower.** ***A.*** Electron micrograph of a bee's head. Its antennae can sense surface textures and air currents, tastes, smells, temperature, and electric fields. The large curved eyes on both sides of the head can look in all directions simultaneously (except to the back) and possess sensitivity to ultraviolet and polarized light. These compound eyes consist of thousands of "micro-eyes" (so-called ommatidia), each with its own hexagonal lens (see inset; scale bar 50 µm), and each of which contributes one pixel to an image. ***B*** and ***C.*** An impression of how a typical star-shaped flower maps onto the bee's curved eyes when viewed from a distance of 4 cm. Note the poor visual resolution and the strongly distorted image from this vantage point.

a completely different significance in the mind of a bee than in our minds. Because plants can turn solar power into an energy drink—nectar—they mean survival for the individual bee and its family. Pollen—plants' sperm—is an equally important resource to collect, since it contains high concentrations of nutritious proteins.

To explore further what plausibly could be on the mind of a creature for whom flowers mean life, imagine a young bee on its first day outside its home. The challenge is to memorize the location of home and landmarks in its surroundings, and to locate profitable flower resources. Moreover, within just a few excursions the bee would be expected to bring home a surplus, or its younger siblings will starve to death. It's clear that our exploring bee must have a large archive of evolutionary knowledge—it does not have to learn to fly, for example, and it has an inborn knowledge that colored, scented dots in the landscape might be flowers.

However, there are many forms of information for which evolution will not have provided the bee with the necessary guidance, since a lot is unpredictable from one generation to the next. The bee does not know from birth where the flowers are, or what exactly they look like; how to manipulate them, whether they contain nectar or pollen, whether they are a good resource or poor; even if they are of good species, they may have already been depleted by competitors. All these things need to be explored and learned by each individual bee. In other words, a bee has to learn a lot in its short adult lifetime of perhaps three weeks, or it will neither find its way home nor become an efficient flower forager.

A bee's first flight is the most dangerous. In bumble bees, up to 10 percent of foragers never return to their native colonies after departing for the first time. Some fail the challenge of accurately remembering the home location; others fall victim to insectivorous

**Figure 1.3. The challenges of being a central place navigator in a natural habitat.** Unlike urban environments (which often contain unique landmarks designed to be recognizable), natural habitats such as forested hills are often replete with repetitive shapes and patterns that offer no particularly memorable features. Yet bees navigate successfully over many miles in such environments, remembering not just the location of their home, but also those of multiple flower foraging patches that can be rewarding at different times of day. Many humans, if forced to operate in such environments without modern technology, maps, or help from knowledgeable guides, might fail such spatial challenges.

birds or sit-and-wait flower predators such as crab spiders. To appreciate the nature of the challenge, imagine human children in this situation. To roughly match the endowment of a few-days-old novice forager bee, let's assume our experimental children are already a few years old (say six years, so of school age). You release them into a wild environment—that is, one without purpose-built, memorable landmarks such as buildings (figure 1.3). Let's make things simpler for the children and keep our environment predator-free. Their only instruction is to bring back food that, like a bee's food, is perhaps up to five kilometers away from home. They need to have the forethought to take sufficient provisions to survive the trip, and when they run out, the resourcefulness to find their own. To match the complexity of floral structures, let's assume that the food needs to be extracted from a variety of puzzle boxes whose mechanics must be figured out by the children themselves, without any instructions from adults. Then, without the assistance of well-meaning passersby, they must find their way home. How many do you think you would see at the end of the day who would also be carrying a significant surplus of food?

It is clear that those few who might succeed would be the ones with extraordinary spatial memory, good searching and motor learning abilities, and fine judgments of the quality of various resources. Over the next few days, some individuals may get better and better: having remembered the most profitable vending machines, they will focus on exploiting these (and recognizing others like them), and will also find shorter paths to connect the best locations. But things won't be entirely stable. Let's introduce some competition from a different group of children, and also some unpredictable changes, as in the flower world: a previously profitable location vanishes and new ones emerge, requiring further exploration. These are just a few of the basic challenges that a bee faces, and that therefore might occupy her mind. In the following sections, we will learn that these challenges require many forms of complicated decision making and efficient memory organization.

## The Mind of a Shopper in the Flower Supermarket

Flowers are, essentially, plants' sex organs, and their colors, patterns, and scents are designed to lure animals into a sexual transaction that many plants, given their lack of mobility, cannot accomplish without help: the transfer of pollen from male flower parts to female ones. But bees don't usually provide this service for free; they need to be rewarded for their efforts. From this perspective, pollination systems may be viewed as biological markets in which animals choose between "brands" (flower species) on the basis of their quality (sugar content of nectar, for example) and plants compete for "customers" (pollinators). Bees learn the advertisements that flowers display and link them to the quality of the product each flower contains. The offerings of this market are constantly in flux: a flower patch that was rewarding in the morning may cease to yield nectar by lunchtime, or may have been depleted by competitors. It may be rewarding again at the same time the next morning, but then have withered altogether three days later. Foraging bees need to update their information in light of these changes and juggle exploitation with prospecting for alternative sources.

Much of the workings of the bee's mind can be understood only when one considers the natural challenges of the constantly changing market economy in which it must operate. The pressures of operating in this setting are often expressed in terms of physical performance. For example, a bee can carry its own body weight in nectar and/or pollen; it may need to visit 1,000 flowers and fly 10 kilometers to fill its honey stomach only once; and 100 such trips may be required to generate a teaspoon of honey. Less appreciated are the mental efforts required along the way:

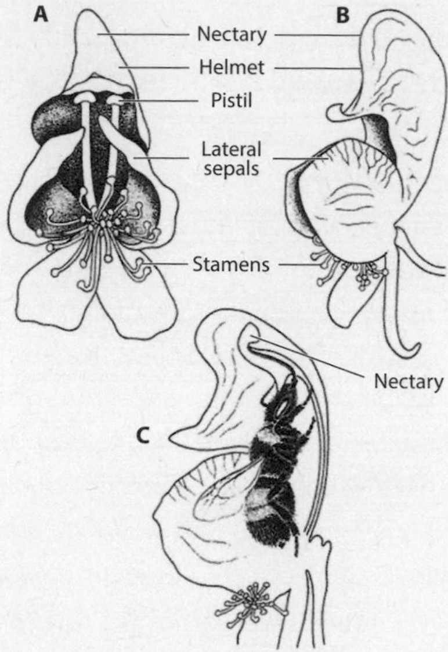


**Figure 1.4. A flower as a natural puzzle box.** ***A.*** frontal view and ***B.*** lateral view of monkshood flower (*Aconitum variegatum*); and ***C.*** A bumble bee inside the flower, inserting its tongue overhead into the "hood" of the flower to extract nectar. Naïve individuals often fail to locate the nectar, and even those that succeed take dozens of visits to hone the technique.

in visiting 1,000 flowers, the bee has to work 1,000 floral "puzzle boxes" whose mechanics can be as complicated as operating a lock (figure 1.4), and no two flower species are quite alike in the mechanics that have to be learned to gain access to their contents. While flying through a flower meadow, the bee is constantly bombarded with stimuli (color patterns, scent mixtures, electric fields) from multiple flowers of several species *per second*, requiring the bee to pay attention only to the most relevant stimuli and to suppress the rest. Between visits to 1,000 flowers, the bee may have to

**Figure 1.5. Shopping in the flower supermarket.** A bee flying over a flower meadow is faced with a bewildering tapestry of sensory stimuli, such as the colors and scents from multiple flower species. Like a human shopper, the bee must identify those flower species ("products") that deliver the best cost/benefit ratio (i.e., the best nectar and pollen rewards after the efforts to reach these rewards are considered). She must memorize these flowers' advertisements (e.g., their color, shape, and scent) and focus her attention on just these flower species, avoiding distraction from other flower signals.

reject 5,000 other flowers that either are unfamiliar or have been found to be poorly rewarding, or only rewarding at a different time of day (figure 1.5).

While foraging, the bee also has to overcome the frustration and the starvation risk of finding dozens of empty flowers in a row that a competitor has recently emptied, and she must decide when to cut her losses and explore for an alternative food source. As she keeps visiting several thousands of flowers a day, rules begin to emerge; for example, are bilaterally symmetrical flower species (such as snapdragons) more rewarding than radially symmetrical ones (such as daisies), irrespective of species and color? Learning rules is not typically regarded as within the reach of an insect mind, but as we will soon discover, the pressures of operating in the flower supermarket have given rise to such intelligent operations in the bee. What's more, while figuring out all these contingencies, she also has to dodge attacks from predators, and remember and avoid flower patches where predation risk is especially high. She has to keep track of the location of her home no matter how convoluted her flight path, and in the face of wind gusts that might displace her far from her established route.

## Complex Decisions, Communication, and Construction at Home

Finally, upon her return home, the bee might find that a bear is digging up her nest. What to do? Should she first unload her forage, or attack the bear and risk being killed? Should she buzz around the bear's head in a menacing way in the hope that this will suffice as a deterrent? Or should she slyly wait on a nearby tree until the attack is over? You might think that these choices are decided by innate priorities, but individual bees can choose differently according to their own predispositions.

After the bear is gone, nest repairs are needed in addition to replacing stolen honey. Building honeycomb requires the construction of precisely hexagonal cells from slivers of soft material that emerge from the abdomen, the cells sized roughly to fit around a bee's body. For unknown reasons, this task involves forming hanging chains of bee workers (figure 1.1). Bees have to hold hands with their sisters, suspended in midair, while repair work continues around the clock.

Night or day, it is always dark in a typical western honey bees' nest (one undamaged by a bear, that is), and the world inside is no less fascinating and strange than a bee's outside world. Imagine a 100-story windowless skyscraper, as packed with people as a bus during rush hour. All main surfaces are vertical, and individuals are constantly scurrying up and down walls. How does any individual know what to do, among the dozens of tasks that need to be taken care of by the colony as a whole?

Much of bee communication works by pheromones (cocktails of chemicals released by multiple glands distributed over the body—fifteen in the case of the honey bee) and by electrostatic signals bees can generate, which they sense with mechanosensory hairs. But honey bees can also communicate about flower locations using symbolic movements: a strange motor display called the dance language. A forager performs a solo dance on a vertical wall. Imagining you are a bee, you and several other bees try to decipher from the dancer's moves where she has found a food bonanza. It's dark, so to read her movements, you have to touch her throughout her dance. A bee does this by putting its feelers on the dancer's abdomen and holding them there while the dancer turns and shimmies. To add an evolutionary perspective, imagine your *life* depended on how well you could sense and interpret the dancer's movements. Some of us would do better on a dark dance floor than others. Some would fail altogether. Some individuals might have a knack for communicating via dances in the dark and be pretty good at it right away; others would be more adept at *learning* to communicate this way. In time, over many generations, there would be selection both for particularly efficient ways of encoding messages in dances, and for aptitude in tactilely deciphering the code.

## Why Imagining Other Minds Is Important for Understanding Them

Some philosophers may argue that there is no point in trying to imagine such strange alternative worlds. I disagree; I think that it's tremendously useful. I cannot precisely imagine what it's like to be you (much less any other animal), but I can do it a little bit, knowing you. I cannot imagine whether you see red the same way I do, but I can check the extent to which you and I agree on calling the same things red, and the extent to which we each discriminate between two similar shades of red (which a bee cannot do). I can also imagine what it's like to have reduced sensory powers (it happens when I take off my glasses, or when I am feeling my way around a dark cellar and have to use my tactile sense to compensate for the lack of vision), and I can imagine, a little bit, what it's like to have a sensory superpower, such as X-ray vision. If I did, it would certainly be accessible to experimentation—for example, you could measure how thick a wall my vision could penetrate, whether I could detect the color of someone's clothes through a wall, and so on. Such tests of what another being can perceive help us to imagine its world a little better.

The question that so puzzles some philosophers—is it unknowable *what it's actually like* to be some other animal?—is probably relatively pointless. After all, it feels entirely unspectacular to live in an alternative sensory world once you get used to it.

Only the transition of acquiring a new sensory ability, if we could do so, would feel exciting; but it would lose its strangeness quite quickly, and then it would feel normal. Sensory perceptions only become subjective *experiences* in any meaningful sense if they are attached to emotional experiences—in a bee's case, perhaps, her reactions to finding a food bonanza, escaping a crab spider attack, or seeing her nest ripped apart by a large mammal. We will learn that bees have mind states that, by the same criteria applied to domestic and wild animals, would certainly qualify as "emotion-like states."

To explore what it might feel like to experience life from any animal's perspective, understanding what is important for that animal, as we have done above, is a necessary starting point. If we grasp that animals such as bees perceive the world through entirely different sensors than ours, and that different aspects of the environment are relevant for their well-being and survival, then we can unleash our imagination, immunized against the risks of anthropomorphizing—inappropriately reading human-like psychology into animals' behavior.

## Which Bees?

When contemplating bees, most of us think of social species, most prominently the domesticated western honey bee, *Apis mellifera*. Indeed, much of what we know about the psychology of bees has been studied in this ubiquitous species and a handful of other social species, such as bumble bees. Their social lives come with fascinating aspects of psychology. For example, they employ highly complex communication systems that ensure the efficient division of labor inside the colony to provide adequate nutrition, climate control, and colony defense. Yet only a few hundred of the world's 20,000+ bee species are social, and the biology and behavior of the many solitary species are no less fascinating. These bees, too, provide for their young and build a home for them—but they are single working mothers, for their males, just like those of social bees, are good only for sex. Solitary females face many of the same learning tasks as social bees—for example, acquiring the spatial memory for home, and learning the appearance and handling techniques of various flowers. Solitary bees face the additional challenges of being "jacks-of-all-trades": while social bees can allocate tasks to groups of specialists, solitary bee mothers must do it all themselves: locate suitable nesting locations, construct nests, defend them from parasites and predators, and provision the brood. However, I will not attempt comprehensive coverage of the literature on the psychology of these many species of bees. Instead, my focus will be on informative examples from across the world of bees.

## Road Map of the Book

The book is structured as follows. This introduction is followed by an overview of the bees' sensory toolkit (chapters 2 and 3). This is important because all the information that is stored in a bee's mind must first pass through its sense organs—and we will discover shortly that the sensory world of bees is not only wholly different from that of humans, but also quite possibly richer. However, not everything that is in an animal's (including a human's) mind is individually acquired: our instincts govern, at least in part, what we desire, what we fear, how we perform certain movements, and so on. Chapter 4 is about the diverse repertoire of bees' innate behaviors, and the extent to which these govern their psychology and learning behavior. Following this, in chapter 5 we explore how the roots of bee intelligence can be sought in their lifestyle as *central place foragers* (i.e., those that have a home to return to). Bees' ancestors had already switched from a vagabond lifestyle to one in which adults built nests where they protected and provisioned their offspring, and this required a keen spatial memory to ensure that the location of the nest could always be found even after long-distance excursions. Chapter 6 contains a detailed exploration of how bees' minds represent space.

In chapter 7 we learn how the habit of flower visitation drove bees to be the intellectual giants of the insect world; how, beyond the basic need to learn flower locations, colors, and scents, bees also learn rules and concepts over their lifetime that help them exploit flower resources efficiently. Chapter 8 turns to bee social learning. Bees can learn a surprising amount of information from observing peers, including which flowers to visit, but also how to solve complex object-manipulation tasks. Accordingly, many complex social behaviors are much more driven by individual problem solving than by a diffuse swarm intelligence, as was traditionally thought.

Having covered the ground from sensory input to complex social cognition, in chapter 9 we explore how the miniature nervous system of bees can support such stunning complexity. Chapter 10 focuses on psychological differences between individual bees, and their neural underpinnings. In chapter 11, we draw on evidence from all the preceding chapters and ask perhaps the most difficult question: Are bees conscious? Given that the answer is most likely "yes," we end in chapter 12 with ethical considerations relating to bee conservation, as they emerge from our explorations of subjective experiences and the likelihood of at least a basic emotional life in bees.

## A Historical Framework

Bees, and the sweets they provide, have been with humans right from the beginning of our evolutionary history. Our closest relatives among the apes consume honey and use tools for its extraction from wild bee colonies. It is therefore eminently plausible that the earliest hominins did the same. Prehistoric cave painters on several continents immortalized the raiding of bee colonies, and the members of many extant hunter-gatherer tribes extract honey from multiple species of wild bees. Honey is the most carbohydrate-rich energy drink that nature has to offer, and some scientists now believe that the practice of efficient honey collection might have fueled the evolution of our energy-hungry brains.

But, as many creative persons will testify, sugar is not all that's required to fuel the generation of bright ideas. And indeed bees provided for inebriation, too: mead, made from fermented honey, is one of the oldest alcoholic beverages. Mead has been consumed for at least 9,000 years, in countries as far apart as China, Finland, Ethiopia, and pre-Hispanic Mexico. And candles made from beeswax lit up the night (and scholars' desks, and temples) for many millennia before the advent of electric light.

Given the long-standing relationship between humans and bees, it is perhaps unsurprising that there is a vast body of scholarly work about the behavior of bees. During research for this book, I have enjoyed browsing the historic literature on the topic, such as the works of the blind Swiss scholar François Huber, who reported, around the turn of the eighteenth to nineteenth century, on the possibility that planning abilities were involved in the comb construction of honey bees, as well as on interindividual "personality" variation, by which he sought to explain labor division in the colony. Another inspiring story is that of the African American scientist Charles Turner (1867–1923), who performed pioneering experiments on the psychology of bees and other insects while working against impossible odds, as a high school teacher without access to scientific laboratories or libraries.

Some of this historical literature is so little known by today's scientists that unearthing it is as exciting as if these discoveries had been made in one's own laboratory. Throughout the book, I have provided a historical context to more-recent findings, and we will see that many seemingly contemporary ideas about the minds of bees had already been expressed, in some form, over a century ago. Since our scientific forebears were often also excellent writers, with styles less dry and jargon-y than many of today's scholars, I will give you tastes of these historic writings, in the hope that this will inspire you to explore the original works. I have also dedicated space to describing at least some biographical details of the scholars whose breakthroughs have inspired me. This is because no scientist operates in a vacuum: it is often important to understand, for key discoveries as well as important errors, the times and circumstances in which these scientists operated and who influenced them.

Come with me on this journey into the minds of bees. We begin with their alien sensory world.

# 2

# Seeing in Strange Colors

> The assumed attractiveness of bright colours to insects would appear to involve the supposition that the colour-vision of insects is approximately the same as our own. Surely this is a good deal to take for granted.
>
> **—Lord Rayleigh, 1874**

To explore what is in the mind of the bee, we first need to understand its senses, because all the information that an animal ever acquires is first filtered through its sense organs—and these differ profoundly between species. We will learn in this and the following chapter that the sensory world of bees is in no way impoverished compared to ours, despite their miniature nervous system. Bees have all the traditional classes of senses that we do (touch, vision, hearing, smell, taste, temperature), and also some of those that don't spring so readily to mind (such as the senses of equilibrium and of time). They also have senses that we lack (including a magnetic compass). But what is striking is how profoundly different the world of bees is from ours in every possible sensory dimension. We begin in this chapter by exploring the color sense of bees, which, as Lord Rayleigh intu-

**Figure 2.1. Bees can see ultraviolet light, and can therefore see patterns of flowers that are inaccessible to human observers.** The petals of the flower shown here are a unicolor yellow for us (***left***), but two-colored for bees, as evidenced by an image generated with a special UV-transmitting filter that blocks all light visible to humans (***right***). The bumble bee's white abdominal region is also UV-reflecting, while the yellow stripes and black regions of the bee's body are not.

ited above, is profoundly different from that of humans. We will use the color vision of bees as a case study in how to explore an animal's senses, before turning (in chapter 3) to the exploration of bees' other sensory modalities.

The first person to experimentally explore the alien color sense of insects was John Lubbock (1834–1913—more of him in chapter 3). Lubbock observed that if he exposed ant colonies to light, they would move their brood from brighter to darker areas. Then he passed the light through various color filters. He found that the ants would move their brood away from violet light, even though this wavelength appeared almost dark to humans. Lubbock noted, "It would . . . seem that their sensations of color must be very different from those produced upon us. But I was anxious to go beyond this, and to attempt to determine how far their limits of vision are the same as ours." He placed the pupae of ant colonies under ultraviolet light—and the workers of various species were quick to remove their larvae from the potentially damaging radiation, wholly invisible to us. Moreover, the ants often shuffled their larvae into red light, which appeared very bright to a human observer, but for the ants seemed close to the complete darkness they were seeking for their brood. This was a first hint, formally confirmed decades later, that many insects are red-blind, or at least that their sensitivity does not extend as far into the long wavelengths as ours.

The discovery that insects are sensitive to a portion of electromagnetic radiation to which humans are blind, opened up a window into sensory worlds entirely different from our own (figure 2.1). We now know that most animals (and all bees) can see ultraviolet light—we humans (and most mammals) are quite unusual in missing out on this sensory dimension.

## The Carl von Hess vs. Karl von Frisch Debate on Bee Color Vision

Working with individually trained honey bees, John Lubbock demonstrated that bees could learn to associate different colors of paper with honey. However, the German ophthalmologist Carl von Hess (1863–1923) pointed out that this was not yet a formal proof of color vision: even a totally color-blind human can distinguish, say, red from blue, since the two are typically different in intensity. Likewise, two differently pigmented papers might appear as different shades of grey to a color-blind animal. In the first comprehensive book on color vision in animals, published in 1912, the eminent von Hess, who had been knighted for his work on the science of vision, concluded that all invertebrates (and also fish) are color-blind.

In the same year, an Austrian assistant professor in his mid-twenties, Karl von Frisch (1886–1982), made the plausible argument that there would be little sense in colored flowers existing if pollinators couldn't see the colors. Why else would evolution make sure that the flowers of most plants stood out in color from their leaves? Von Frisch developed an experimental paradigm that proved von Hess wrong. He gave honey bees a sugar solution reward in a small glass bowl set on a colored rectangular or square card, set in an array of other cards in various shades of grey (figure 2.2). The bees invariably located the colored card among the grey ones, even when the spatial position of the colored card was shuffled (so the bees could not simply have remembered the position of the target).

In the German academic system of that time, disagreeing with an established professor could spell the end of a young scientist's career. And, as one might predict, von Hess was infuriated. Having caught wind of von Frisch's experiments, he rushed to publish his own account before von Frisch could even get his work into print. Von Hess used honey as a reward in his attempts to train bees to learn colors (rather than the scentless sugar water von Frisch had chosen), but honey has such an attractive scent for bees that it can override any attempt to train them to other target features. Von Hess's results were negative. In his 1913 paper, titled "Experimental Investigations about the Alleged Color Sense of Honey Bees," he blustered: "It was possible to demonstrate that the older claims of Lubbock . . . , as well as recent ones of von Frisch, according to which bees can be 'trained' to certain colors, are wrong altogether. . . . Not a single fact is known that could even make plausible the notion of a color sense in bees. . . . Through my investigations this assumption is terminally refuted."

Von Frisch was neither impressed nor intimidated; in his detailed accounts of his experiments, published in 1914, he explained

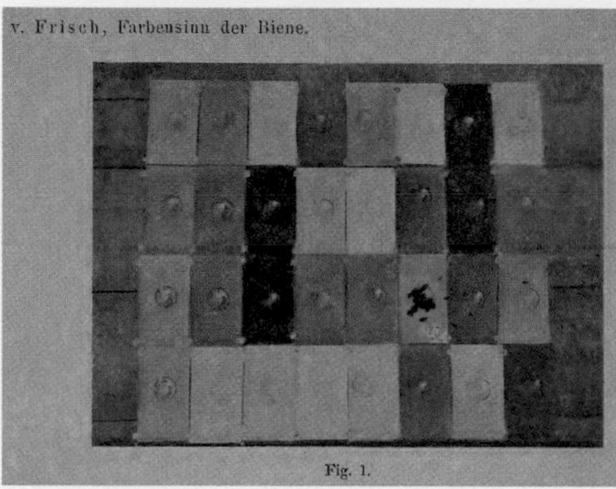


**Figure 2.2. Original color figure from Karl von Frisch's pioneering 1914 paper demonstrating color vision in bees.** Bees trained to collect sugar water from glass dishes located on blue cardboard correctly found "blue" even if it was presented in a new location, and could distinguish it from all shades of grey, thus showing that they had not simply learned the brightness of the correct stimulus.

the evidence for the color vision of bees carefully and conclusively. Commenting on von Hess's writings, he did not mince words, either:

> Of course, von Hess does not admit this. He tries to discredit my work, again and again, by claiming that it is amateurish and performed without any knowledge of physics or the physiology of colors. No conclusive proof is delivered for these claims. . . . He thinks all my decisive experiments are erroneous. . . . I protest against this . . . polemic approach, and I request that von Hess refrains from using such contemptuous adages. . . .
>
> Von Hess . . . presented his bees, that he had trained to blue, with a yellow pencil sullied with honey and saw that they landed

> on it; on a blue jacket they landed only after he had sullied it, too, with honey. These experiments only show—as everybody knows—that bees can be tempted with honey.

As a young scientist, von Frisch had put his future on the line by publishing such statements. He wrote to his mother, "I have the uncomfortable feeling that I now have a real enemy in the world out there, the first one, and someone who could really damage me." But von Frisch's evidence was so overwhelming that von Hess's attempts to discredit the young scientist failed. In his autobiography, von Frisch would later acknowledge that the altercation had actually strengthened him by giving him wide exposure. It certainly prepared him for defending future discoveries with irrefutable evidence and arguments. Von Frisch went on to win the Nobel Prize in 1973, while von Hess's views on the topic sank into obscurity.

Incidentally, von Hess was actually correct about bees' unresponsiveness to color in one particular experimental paradigm he had tried: *phototaxis*, or the attraction to light, which many flying animals display when under threat. It turns out that this behavior is indeed color-blind in bees. But being color-blind in one context does not mean that the organism can't see color. All humans, for example, are color-blind in dim light or darkness—hence the expression "At night, all cats are grey." But both humans and bees see flowers in color, at least during the day. Von Frisch also found further confirmation for the notion that bees' color vision is profoundly different from that of humans: he discovered that while bees had no difficulty locating a blue or yellow square among the grey squares, they would invariably confuse red with the darkest greys. He concluded that bees are red-blind, which explains why red flowers are relatively rare in the European flora.

## Karl von Frisch and the Nazis

Karl von Frisch then turned his attention to other subjects (most notably the exploration of the bee dance "language," which we will discuss in chapter 5). UV sensitivity in a number of bee species, and UV reflectance of flowers, were discovered in the 1920s, and von Frisch left the in-depth exploration of color vision to his students, most notably Karl Daumer in the 1950s (see next section). But these further investigations might never have materialized, because von Frisch came under attack during the Nazi era, 1933–1945. He had a grandmother who, though baptized as a Catholic shortly after birth, had had Jewish parents. The Bavarian State Ministry demanded that von Frisch, a "second degree mongrel" (their wording in a letter dated January 12, 1941), be removed from his university post in Munich. Influential colleagues accused him of "bigoted opposition towards antisemitism," and one of his publications was denounced as "a prime example of Jewish propaganda." They lamented that "the most modern and best-equipped German zoology institute is currently governed by a small-minded, narrow specialist, who is uncomprehending and hostile towards the new epoch. It is high time that the institute obtains a leadership that terminates these conditions."

Perhaps in desperation, von Frisch appears to have gone some way to pander to the Nazis (though he never became a party member): his otherwise insightful 1936 book *You and Life: A Modern Biology for Everybody* contains disturbing final paragraphs on "racial hygiene" and the recommendation to sterilize mentally retarded people without consent. In these paragraphs, von Frisch, himself so severely nearsighted that he had been deemed unfit for military service in World War I, laments the fact that modern civilization mollycoddles the nearsighted, who would never have survived our primordial ancestors' harsh struggles for survival. One hopes that these passages were added under pressure from Nazi authorities, and not volunteered by the man himself. Perhaps by going to some length in what was requested by the authorities, von Frisch hoped to protect not just his own job, but also the multiple Jewish researchers still working at his institute in the mid-1930s. In any case, the text prompts uncomfortable thoughts about how an intellectual, himself at risk from Nazi ideology, tried to navigate those extremely difficult times.

In the end, von Frisch's position was temporarily rescued by a honey bee disease outbreak. In 1940–1942, *Nosema*, a single-celled gut parasite, wiped out several hundred thousand beehives (an early "colony collapse disorder" epidemic), causing significant threats to food security; many crop plants remained poorly pollinated. Martin Bormann, head of the Nazi Party chancellery and a close personal friend of Hitler's, directed that von Frisch's dismissal should be postponed to the end of the war (presuming that the Nazis would win the war and sack von Frisch after the victory). Von Frisch took on the thankless task of honey bee disease management until 1945. He was unsuccessful in finding a treatment for *Nosema*, but the redeployment saved his job and his bee research, and gave him the opportunity to train more young scientists. He went on to make many more groundbreaking discoveries, and most of the scientists whose work is portrayed in this book are his students, or his students' students, or his students' students' students (I'm one of them, but I'm just one of many).

## A Different Colored World

Karl Daumer (b. 1932), another student of von Frisch, discovered both similarities and differences between human and bee color vision. His PhD work made the bee the best-investigated animal for its color vision besides humans, an eminence the bee has retained ever since thanks to several generations of talented researchers who followed in his footsteps. Before we explore how bees see colors, a few words about *our* color perception.

Color vision is a little strange even in humans, in that the perception of color does not easily allow us to reconstruct the physical spectral properties of an object. If we mix yellow with red light, we will see orange—and not only will we be unable to tell that the color has been produced by mixing two lights, we will also be unable to distinguish the mixture from monochromatic (single-wavelength) orange light. The perception of white can be generated by mixing any pair of complementary colors—blue and yellow, red and cyan, green and magenta—or the three colors of light that are "primary" in our visual system: green, red, and blue.

By combining the short-wavelength and long-wavelength ends of the spectrum (violet and red), you generate a perceptual quality that does not exist in the spectrum at all: purple. We are so used to these mixture phenomena that most of us are unaware of how poorly they correspond to the physical world: the physical stimulus (or stimuli) simply cannot be deduced from any single color we see. In contrast, in hearing, you can perfectly well distinguish two tones, or indeed a chord of three tones. We would never perceive a mixture of 400 Hz and 800 Hz as an intermediate frequency (say, 600 Hz)—but that is exactly what happens in vision.

The difference begins with how the receptors are organized. We have only three color receptor types (for blue, green, and red light), and every one of the million or so colors that we perceive is generated by the relative stimulation of these three receptor types. In contrast, in our inner ear, we have thousands of auditory receptor cells responding to different frequencies, and their responses are processed in parallel, rather than competing against each other, as in color vision.

The bee visual spectrum is shifted in its entirety to a range of shorter wavelengths than humans', from about 300 nm (ultraviolet) to 650 nm (yellow-orange; figure 2.3). Having built a sophisticated apparatus for mixing monochromatic lights, Karl Daumer discovered that honey bees are more sensitive to ultraviolet than to any other color, and that similar mixture rules hold in bee color vision as in human color vision (figure 2.4). You can mix two monochromatic lights (say, blue and green) to produce a mixture light that is indistinguishable from a single intermediate wavelength (turquoise). As in humans, the short- and long-wavelength ends of the bee visual spectrum can be mixed to generate a unique sensation that cannot be generated with any single wavelength (Daumer called this "bee purple"). He also found that there are complementary colors in bee vision, such as blue-green and UV, violet and green, "bee purple" (UV plus green) and blue. Bees confuse white surfaces that do not reflect UV with blue-green (turquoise).

Daumer concluded, based solely on psychophysical experiments and without any direct access to a bee's brain, that bee vision, like that of humans, must be *trichromatic*—that is, have three principal variables, in the bee's case based on UV, blue, and green receptor signals. Up to that time, however, this trichromacy theory was just that—a theory—both for humans and for bees. In 1962, for the first time in any animal, the German physiologist Hansjochem

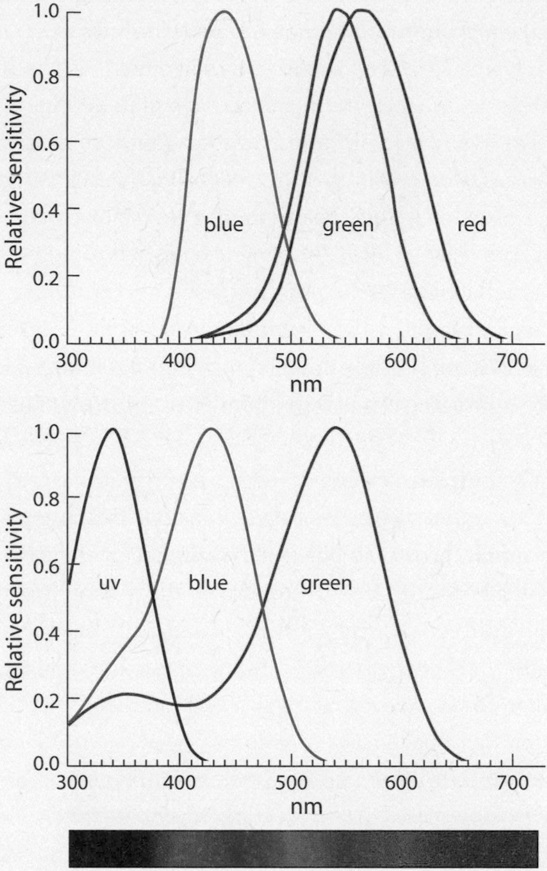


**Figure 2.3. The color receptors of bees compared to humans.** The spectral sensitivities of the color receptors of humans (***top***) and bee color receptors (***bottom***) on the wavelength scale from 300 nm to 700 nm (ultraviolet to red). Both species have three color receptors, each with a peak sensitivity at a certain wavelength, and sensitivity falling off to both sides of this peak. While humans are entirely insensitive to UV, bees have a specialized receptor in this region of the spectrum. On the other hand, the sensitivity of the bees' long-wavelength receptor does not extend as far into the red as that of humans.

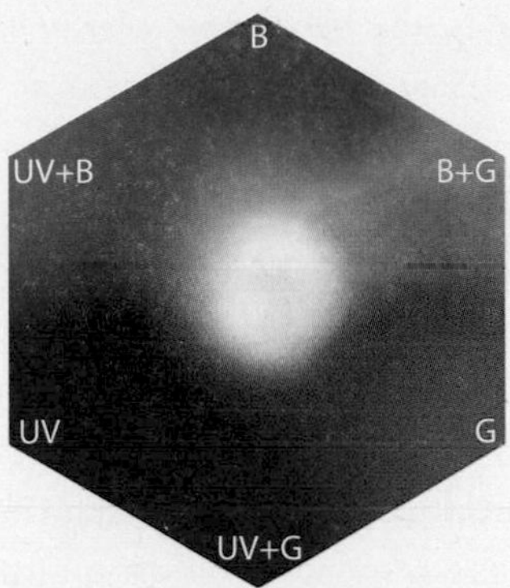


**Figure 2.4. The color space of bees illustrates the principles of color mixture.** The angular direction (as measured from the center) informs us about the bee-subjective hue. Objects whose reflectance primarily stimulates one photoreceptor (UV, blue, and green) will be in the bottom left, top, or bottom right corner, respectively. Mixtures (such as between green and blue, in the top right area) can be anywhere in between. The bee color space also contains a region that (like humans' purple) is not present in the spectrum—namely, a mixture between the long-wave (green) and the short-wave (UV) ends of the spectrum—in the bottom region of the color space (called "bee purple" in some articles).

Autrum (1907–2003), Karl von Frisch's successor at the University of Munich, and team members managed to insert microelectrodes (glass capillaries with a tip diameter of 1/10,000 mm) into the tiny, single photoreceptor cells in bees' eyes, and recorded their electrical signals while pointing lights of different wavelengths at the bees' heads. They confirmed that honey bee eyes contain three types of spectral receptor cells, one with maximum sensitivity in the green, one in the blue, and another in the ultraviolet. The receptors were each sensitive to quite a broad range of wavelengths around their peak sensitivity curves, which are roughly bell-shaped (figure 2.3).

Randolf Menzel (b. 1940), a German neuroscientist and academic "grandchild" of von Frisch, measured how quickly honey bees could learn to associate various colors with sugar rewards. It turned out that they were exceptionally fast. John Lubbock had already suggested that blue was a honey bee's favorite color, and indeed, Menzel found in his PhD studies that just one reward on a blue-violet color was sufficient to establish a highly accurate memory. Bees would subsequently pick such colors with high accuracy over any others that were presented. Most other colors took a few more sugar rewards for the bees to form durable memories, but even unpopular colors, such as turquoise, would be learned after ten sweet rewards.

No other animal learns colors this rapidly. In the decades since these groundbreaking studies on honey bees, color learning has been tested in many more animals. In a comparative analysis of color learning speed in eleven animal species, honey bees were the fastest, followed by fishes, then birds . . . and the slowest were human infants. Curiously, this result, which runs counter to the expectation that humans are "smartest," was then used in a textbook on animal learning to argue that learning speed is not a useful measure of intelligence. There may be good reasons not to equate learning speed with intelligence, but the fact that humans do not top the chart should not be one of them. The reason bees perform well at color learning tasks is that evolution has prepared them to memorize floral cues: worker bees are natural-born color choosers who constantly evaluate the flower offerings in their flight range, and must rapidly learn if one color is no longer rewarding and another color of flower is becoming richer in nectar or pollen offerings.

## Did Bee Color Vision Evolve in Response to Flower Colors?

An obvious question is why the bees' color vision is so different from ours—why, for example, they see the ultraviolet, and why their photoreceptors are tuned to the exact wavelengths they are. A likely answer is that this must be related to their habit of flower visitation, and to the particular colors that flowers display to bees.

In my PhD studies (1991–1993), I was in the right place at the right time to put this hypothesis to the test: my supervisor, Randolf Menzel, had just started collaborating with the pollination biologist Avi Shmida, and during a field trip to Israel, they had measured the physical color properties of dozens of flower species. This involved measuring the amount of light bounced back from the flowers at every wavelength to which animals can be sensitive—from 300 nm (ultraviolet) to 700 nm (far red). Because it was not possible at the time to genetically engineer bees with modified color vision, an alternative approach was to model various color vision systems, some like that of the bee and some completely different, and then to ask what would theoretically be the best system to detect and recognize actual flower colors.

I essentially fed all these reflectance functions into a computer and asked the machine to find the optimal color vision system. I varied everything—the wavelength positions of all three color receptor types (normally sensitive to UV, blue, and green in bees), the neuronal processes evaluating the receptor signals, and the illumination conditions under which the flowers presented themselves to the bees. The results of any simulations often

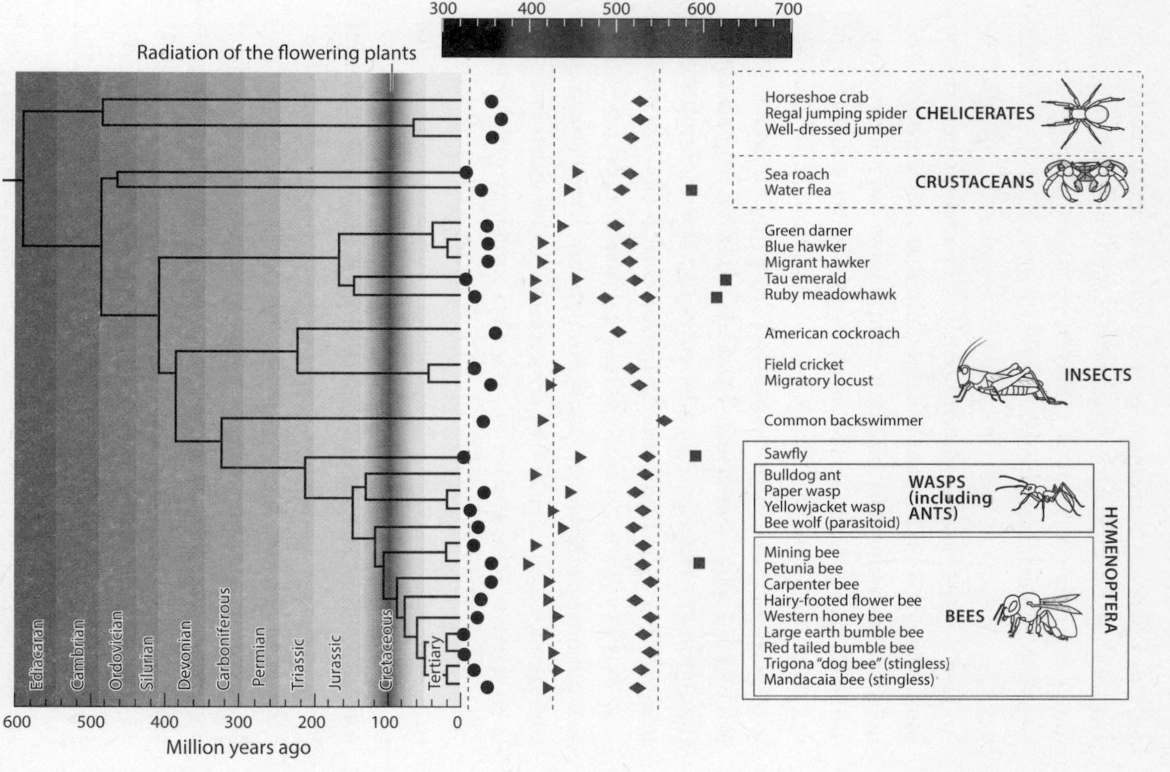


**Figure 2.5.** (See facing page for figure title and caption.)

appeared only several days later—the available computers were slow, and my basic BASIC programming skills didn't help—but in those days, no convenient software packages were available, so every modeler had to tailor-make their own software from scratch.

The results were ultimately very striking, however. These computer-generated optimal color receptor sets were almost identical to the ones in real bees' eyes (figure 2.5). While many qualitative observations had suggested that sensory systems are matched to an animal's ecological niche, this was the first quantitative demonstration that an animal's color sense is precisely optimal for the task at hand. An engineer could hardly design a better color vision system for coding flower colors than the one implemented in bees. Using similar modeling approaches, researchers later explored the adaptive significance of many other biological sensors, for example, primate color vision and its relationship to fruit colors. I was very excited by this apparent tuning between the color vision of bees and the colors of flowers. So were many of my

◀ **Figure 2.5. Values of maximal sensitivity of the color receptors of representative arthropods (insects, crustaceans, and chelicerates) and their phylogenetic tree.** The tree illustrates how related animal species are, and the approximate times at which they appeared on Earth. Black circles: positions of maximum wavelength sensitivity of UV receptors; triangles: blue receptors; diamonds: green receptors; and squares: red receptors. Similar sets of UV, blue, and green receptors appear in almost all arthropods. Red receptors pop up repeatedly in various arthropods, for example water fleas, dragonflies, and also some Hymenoptera. UV receptors were already present in the Cambrian ancestors of the arthropods, and therefore predate the evolution of flower colors by some 400 million years. Black dashed lines: optimal wavelength positions of the photoreceptors for coding flower color.

colleagues—it is the kind of neat story about mutual evolutionary adaptation that everyone likes to see.

But does this optimization of bee color vision to detect and identify flowers mean that the color vision of bees co-evolved with floral coloration? Does the evolutionary influence in this communication system between bees and flowers go both ways? To prove that flower signals indeed drove the evolution of bee color vision, it must be shown that the ancestors of bees possessed different sets of color receptors before the advent of the flowering plants. But how can we determine in what colors insects saw the world 200 million years ago, when flowers first started to appear on the planet?

Evolutionary biologists have a magnificent tool to open windows into the distant past: comparative phylogenetic analysis. For example, all extant mammal mothers feed their offspring with milk. Hence we can deduce with certainty that the Jurassic ancestor of all modern mammals already had mammary glands and used them for the same purpose. In a similar way, we can infer that this ancestor was warm-blooded. In other words, even lacking fossils of morphological details such as glands, we can make inferences about such traits, and even about ancestral animals' behavior and physiology. With respect to our question about bees and flowers, one has to evaluate arthropods (spiders, crustaceans, and other insects) whose evolutionary lineages diverged from those of bees before there were flowers. If the color vision of such animals is indistinguishable from that of bees, this implies that it was already present in a common ancestor that predated the evolution of flower color. Fortunately, comparative physiologists had already collected a large database on color receptors in various arthropods. We needed to map them onto the phylogenetic tree of all the species in question, and any patterns of adaptation should be immediately apparent.

Such a phylogenetic analysis revealed that the kind of color vision that bees have is several hundred million years older than the first flower (figure 2.5). It appears that pretty much all insects have UV receptors, for example dragonflies, locusts, and cockroaches. None of these are your typical flower visitors . . . and even many marine crustaceans have UV receptors. As for the fine-tuning of color receptors along the wavelength scale, there is some variation between species—but no coherent evidence that the habit of visiting flowers has induced a significant change in the color vision of the insects who do so.

We concluded that the Cambrian (and most likely aquatic) ancestor of all insects and crustaceans already possessed UV, blue, and green receptors. Insects were pre-adapted for flower color coding hundreds of millions of years before there were any flowers—before the extensive radiation of the flowering plants that started in the middle Cretaceous (100 million years ago), though perhaps they had their beginnings in the Triassic (250–200 million years ago). In summary, the answer to the question "Why do bees have UV receptors?" is "Because their ancestors did."

The hypothesis that insect color vision evolved in adaptation to particular classes of objects—in bees' case, flowers—must be rejected. At best, having a combination of UV, blue, and green receptors might have been a more general adaptation, useful for coding all sorts of natural objects under varying illumination conditions. Flower colors adapted to insect color vision, not the other way around. In this sense, insect pollinators painted the world. Before plants appointed hungry insects as pollen carriers, the terrestrial living environment was largely green (leaves) and brown (tree bark).

It was through work on the sensory systems of insects that we first came to appreciate that the world we perceive is not an objective, veridical representation of physical reality. It is, rather, filtered through the sensory mechanisms each animal species has acquired during evolution. Now that we have had an introduction to how strangely differently bees see colors, we are prepared to encounter even stranger ways of perceiving the environment in the bee sensory toolkit—to explore sensory modalities that in some cases have no equivalent in humans.

# 3

# The Alien Sensory World of Bees

We find in animals complex organs of sense, richly supplied with nerves, but the function of which we are as yet powerless to explain. There may be fifty other senses as different from ours as sound is from sight; and even within the boundaries of our own senses there may be endless sounds which we cannot hear, and colors, as different as red from green, of which we have no conception. . . . The familiar world which surrounds us may be a totally different place to other animals. To them it may be full of music which we cannot hear, of color which we cannot see, of sensations which we cannot conceive.

**—John Lubbock, 1888**

If Britons today recognize the name John Lubbock at all, it is mostly because he was the father of "bank holidays"—official holidays on which banks and other businesses are closed. In addition to traditional holidays such as Christmas and Easter, the parliamentarian and banker Lubbock introduced several new ones clustered in the spring and summer months. This was because he was also a keen entomologist who lamented that his "parliamentary duties . . . have absorbed most of my time just at the season of year when these insects can be most profitably studied." Of all the ways politicians could use their powers to serve themselves and their public at the same time, letting an entire nation off work so one could study insects is surely the most wonderful.

This son of a wealthy banker had a passion for insects even before he was old enough for school, but his interest took a decisive turn at age eight, in 1842, when his father came home with an important announcement. Little John guessed that he was going to get a pony, but his father explained, "It is much better than that. Mr. Darwin is coming to live at Down." "Down" is, of course, Down House, where Charles Darwin would live for the rest of his life. The house was in the same village as the Lubbock residence.

John Lubbock became "Darwin's apprentice," and between them they undertook some remarkable explorations of the sensory capabilities of invertebrates. Darwin used a piano to test the sense of hearing in earthworms; Lubbock played the violin to bees (in both cases the audience was largely unresponsive). When Alexander Graham Bell traveled to London in 1878 to demonstrate the newly developed telephone to Queen Victoria, Lubbock instantly tested the new technology on ants—to see if they could transmit an alarm message from one nest to another by acoustic means. Thus, before the telephone was used for such mundane purposes as mass communication, it was put to good use in an entomological experiment. The result was negative—but this ultimately led to Lubbock's discovery of a "chemical language" in ants, and the study of pheromone communication in social insects was born.

Some of the results of such unbridled creativity as John Lubbock's are bound to be disappointing (ants won't ever be inclined to use the telephone), but his work revealed many fascinating phenomena. He experimented with intoxicated ants, and discovered that while nestmates would typically help their drunken "friends," strangers from other colonies would unceremoniously dump the inebriated ants into water to drown (an early indication that social insects have strong family bonds and display "kindness" only to kin). But it is in discovering differences between animals' sensory systems that Lubbock provided the most wide-ranging inspiration to the field.

We now know that there are many such differences between humans and insects. For example, while insect spatial resolution is poorer than ours (they see fewer pixels), they can see a lot *faster*—taking in more information per unit time. Typical ceiling strip lights fed by alternating current turn on and off 50 or 60 times per second—invisible to our slow eyes, but literally a stroboscopic light environment for many insects (including bees), which have a visual processing speed five times faster than humans.[5] People often express surprise that insects, with their relatively short lives, can have such a rich psychological life. But their lives are not only shorter, but also *denser*: if the smallest time units one can perceive are constrained by how fast one can see, bees could sense many more events in an hour than humans do.

Sense organs can be in strange places on various kinds of insects. Organs of hearing may occur on the thorax, abdomen, legs, or mouthparts, and some insects are completely deaf (as John Lubbock had suspected when his ants failed to transmit and receive an alarm message between anthills by telephone). Some male butterflies have light receptors in their genitals, which help them find their target during copulation. Having first focused specifically on the bee color sense, we will now explore the other "alien" (to us) sensory modalities of bees.

## Martin Lindauer and the Bees' Time-Compensated Sun Compass

One of Karl von Frisch's outstanding disciples was Martin Lindauer (1918–2008); he, in turn, mentored Randolf Menzel, who mentored me. I also had the good fortune to interact with Martin Lindauer extensively while working at the University of Würzburg in the late 1990s. At the time, Lindauer had formally retired and was marred by advanced Parkinson's disease, but he continued to perform occasional experiments on bee behavior and gave freely of his advice to young scientists. One of fifteen children of a farmer's family, Lindauer had grown up in poverty in the Bavarian Alps. Contrary to the pomposity displayed by so many German professors of his time, and in spite of his superstar status in the field of behavioral ecology, Lindauer remained a humble man all his life. It was fascinating to listen to his accounts of the historical events that he had witnessed, including the history of the science of animal behavior.

Unlike most of his classmates, Lindauer had refused to join the Hitler Youth, but was drafted into the mandatory Reichsarbeitsdienst ("Empire Labor Service") days after finishing high school in 1939 and was forced to dig ditches at Dachau, where the Nazis had established the first concentration camp. Because of his record as a Hitler Youth objector, he was constantly bullied by superiors and peers, and this continued when Lindauer was conscripted into the army in the first days of World War II. He was severely injured in an ambush on the Eastern Front in 1942 and diagnosed as unfit for further army service. The injury most likely saved his life: his company was sent on to fight in Stalingrad only weeks later. Of the 156 men in the company, only three returned from the war alive.

While Lindauer recovered from his injuries in Munich in early 1943, he stumbled into a lecture by Karl von Frisch. He described the experience as something of a revelation: he recounted that between the surrounding sea of primitive lies and brutality and the scientific endeavor of finding objective truth, as conveyed by von Frisch, the contrast could not have been sharper. Even though the battlefronts were still far from Munich at that time, the ugliness invaded the university world relentlessly, too. Lindauer was unaware at the time that von Frisch's post was at risk because of the Nuremberg race laws, but he did experience firsthand the commotion at the University of Munich over political leaflets by a student resistance group called The White Rose, and the subsequent inspection of all students' bags by the Gestapo to check if anyone had so much as dared to pick up any of the leaflets. The authors of the leaflets were betrayed and arrested on February 18, 1943; some of the group, including the youngest, twenty-one-year-old Sophie Scholl, were executed by guillotine just four days later—and several others, later in the same year.

As the Allied bombs came raining down on Munich in the last years of the war, Martin Lindauer began his studies with Karl von Frisch. The zoology building of the university lost floor after floor to the air raids, and after the building was finally completely destroyed, von Frisch moved his center of operations to Austria, while Lindauer stayed behind in Munich and initiated his experimental thesis work in a suburban garden, more or less cut off from direct interaction with his supervisor. Days before Germany's unconditional surrender in May 1945, American tanks rolled right past twenty-six-year-old Martin Lindauer, his experimental setup, and his honey bees.

When the Allies sealed the border between Austria and Germany after the end of the war in 1945, Lindauer became even more isolated from his supervisor. But this may have been useful for building his independence in asking scientific questions and seeing projects through. Students of scientific giants sometimes find it hard to "step out of the shadow" of their supervisors, who often have an endless supply of their own ideas for their team members to study, and sometimes view their team as basically technical support, expected to contribute bricks to build the supervisor's monument. Through the isolation enforced by postwar political events, Lindauer found his own research style and voice early on—and so he was to continue his scientific work with von Frisch more or less as peers for over a decade.

In the 1920s, the German biologist Ernst Wolf (1902–1992) had discovered that bees use a sun compass for navigation. Realizing that he needed a large testing ground that was free of distracting landmarks, Wolf obtained permission to work on an airfield that had been closed at the end of World War I, formerly used by airship-building company Schütte-Lanz. So where 200-meter-long airships had departed and landed just a few years earlier, honey bees were now tested for their navigation abilities.

Wolf discovered that honey bees memorized homing vectors (directions and distances) relative to the sun, so that if they knew the hive was, say, 150 meters south of a feeding station, they would fly 150 meters south even if they had been displaced to a different release site before departing from the feeder. In other words, bees followed a memorized vector, using the sun as a reference. Simply by observing fragments of bee flight, and measuring flight times, Wolf deduced that the flight in such displaced bees comprises three phases: first the straight vector flight, then a searching phase when the bees found themselves in an unexpected location and sought out familiar landscape features, and finally a more or less straight flight to the actual home (figure 3.1).

Using a sun compass is not as easy as using a magnetic compass: Earth's magnetic field doesn't change in the course of the

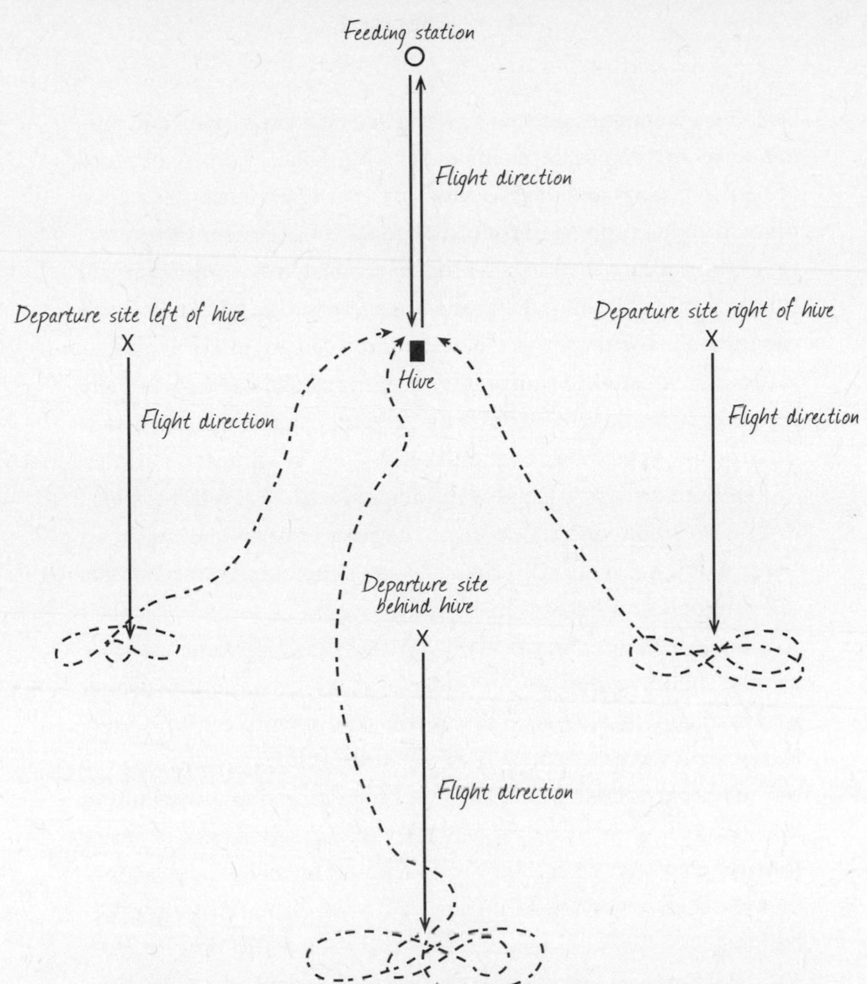


**Figure 3.1. A bee orientation experiment on an abandoned German airship test field.** Ernst Wolf (1927) trained honey bees to shuttle between the hive and a feeding station (top of map) 150 m from home. He then caught experienced foragers when they had filled their honey stomach and were about to head home, and displaced them to three release sites (150 m to the west, south, and east of the hive). Bees displayed an initial "vector flight" (flying the distance and direction they would have flown had they not been displaced), a search phase when they noticed that the vector flight had not taken them to the expected destination, and finally a flight in the correct direction home.

day, and a magnetic compass needle always points north. Using the sun as a compass requires knowing what time of day it is, since the sun's azimuth (the point where a vertical line from a celestial object hits the horizon) does change over the course of the day. After World War II, Martin Lindauer suggested to von Frisch that they expand the exploration of the sun compass. Lindauer wondered what might happen if bees were trained to collect sugar water from a feeding station south of their hive *in the afternoon* (to the left of the sun), and then the hive was moved to a new location—where all the local landmarks were unfamiliar—and presented with feeders in all four cardinal directions *in the morning*. Would the bees continue to fly to the left of the sun (as they had the previous afternoon), or could they time-compensate for the movement of the sun and fly straight south?

Von Frisch was skeptical ("They will think we're insane if we ascribe such complex orientation systems to bees"), but Lindauer persisted, and finally the displacement experiment was performed. Remarkably, bees flew in exactly the same southward direction as they had before displacement—even though the time of day they'd been trained to visit the feeder at the old location was different from the time of day when they were tested. The ability to use a time-compensated sun compass has since been discovered in many other animals.

Lindauer was aware that the usage of a sun compass in bees is far from trivial. Unlike migratory birds, bees do not typically have innately preferred flight directions depending on the time of year. They have to *learn* the route to and from a good flower patch relative to the sun, and the angle between it and the sun's azimuth changes depending on the time of day. Thus they need a sense of time *and* implicit information about where they are on the planet. Either they must have learned it, or the knowledge must be inherited and specific to local populations. Lindauer later took this displacement experiment to an extreme by shipping honey bees from the Southern Hemisphere (where the sun is in the north at midday) to the Northern Hemisphere (where it's in the south at noon)—and in that case, the bees switched their preferred orientation by 180 degrees, as predicted. However, Lindauer discovered that bees actually learn to predict the sun's daily course, and subsequently use the sun compass in the new location correctly.

## Polarization Vision in Bees

Von Frisch and Lindauer understood that the sun's position itself is a very undependable compass cue. The sun is often hidden behind clouds, mountains, or trees—and then how does one know any directions at all? Lindauer recounted to me the sensation of "feeling blessed" when he assisted von Frisch in an experiment that deprived honey bees of a direct view of the sun. Bees still knew the correct direction even when they saw only tiny fragments of the blue sky. Lindauer, and presumably von Frisch, immediately understood that they had tapped into a remarkable ability: bees could reconstruct the position of the sun from patches of blue sky when the sun was obscured. But how?

Von Frisch needed to consult physicists in search of an explanation. They suggested that bees might be able to see polarized light. You may remember from high school physics that light has wave properties, and that light waves "vibrate" in planes at right angles to the direction they're traveling in. Before sunlight hits our planet's atmosphere, the waves all swing in random directions. But inside the atmosphere, light is scattered by air molecules in such a way that in certain sectors of the sky all light waves align in the same direction; the light is then said to be *linearly polarized*. The

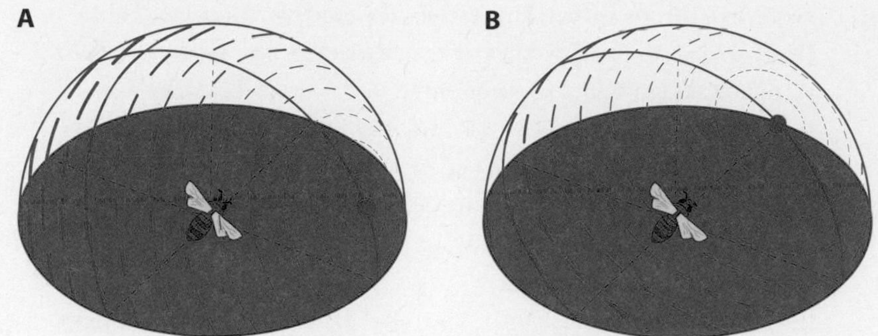


**Figure 3.2. Natural polarization pattern of the sky dome.** Shortly after sunrise (***A***), and when the sun (orange dot) is higher up later in the day (***B***); the bee is at the center of the "flat earth surface." The entire pattern of polarization orientations changes in a predictable manner with the position of the sun. The polarization pattern is essentially arranged in concentric rings around the sun, with the strongest degree of polarization at 90 degrees from the sun, and the weakest degree right around the sun (as indicated by the strength of the bars). The direction of the bars indicates the direction in which the light is polarized.

natural polarization pattern of the sky changes in a predictable manner as the sun moves: close to the sun, the percentage of linearly polarized light is low; it increases gradually to a maximum at 90 degrees from the sun, then decreases again (figure 3.2). So to locate the zone of maximum polarization, point your left arm at the sun, and point your right arm at the sky at a right angle to the left one. Then rotate the right arm while keeping it at a 90-degree angle to the left arm pointing at the sun. The circle you now describe marks the zone of strongest polarization, and the direction of polarization of light is aligned with the circle (figure 3.2).

This pattern can be visualized with an inexpensive linear polarization filter sheet. Such a filter transmits only light waves that swing in a certain direction—rather like a sieve with parallel slits. If you dropped some needles into such a sieve, they would only fall straight through if they were oriented in the same direction as the sieve's slits (all other needles would stay in the sieve). If you hold such a filter at the circle of strongest polarization in the sky, it will let a lot of light through when its "slits" are aligned with the direction of polarized light—the "swinging direction" of the light wave. If you rotate the filter sheet 90 degrees, it appears dark, since all the light waves are perpendicular to the direction the filter allows through. If you hold the filter at the region in the direction of the sun, where the sky's light is not linearly polarized, then rotating the filter won't make much difference in how much light it transmits.

Von Frisch and Lindauer experimented with such polarization filters on their bees. It turned out that just as you can see the sun and infer the invisible circle of most strongly polarized light in the sky, a bee can reconstruct the position of the sun even when it is obscured, because she is able to detect the polarization of light in other parts of the sky. In fact, because the entire polarization pattern of the sky moves with the sun, even small patches of open sky allow a bee to reconstruct the position of the hidden sun—and thus to retain her sense of direction. The experimenters fooled bees with polarization filters, presenting them with patches of sky with an altered polarization pattern. For example, by filtering a patch of sky that contained naturally unpolarized light, bees under the filter would see it as linearly polarized; the bees' sense of direction was thwarted accordingly.

Using the polarization pattern of the sky, bees are able to augment their sun compass with a sensory ability that is almost unimaginable for us humans. It has since been revealed that many invertebrates (including spiders, crustaceans, and all insects so far tested) have sensitivity to polarized light, and even some vertebrates, such as birds, do.

What mechanism endows bees with this ability that we humans entirely lack? This question was pioneered by the German biologist Rüdiger Wehner, one of Martin Lindauer's students. The answer is hidden in the structure of their photoreceptors, the cells in their eyes that convert light into electrical signals. In vertebrates, these cells are called "cones" (for daylight vision) and "rods" (for seeing at night), and the names of these cells give away their shapes. In line with this, insect photoreceptors should really be called "toothbrushes," but they are somewhat less intuitively named *rhabdomeric receptor cells,* and the transparent, threadlike structures that protrude from them like the bristles of a toothbrush are endearingly called *microvilli* (figure 3.3). The sensitivity to polarized light arises because of the particular arrangement of the light-sensitive molecules in these microvilli.

The small diameter of the microvilli (~0.1 μm) forces all the photosensitive molecules (several hundred to thousands per microvillus) to align with the parallel axes of the microvilli (tens of thousands of them per receptor cell)—so for each photoreceptor "toothbrush," the molecules all point in the same direction. When the e-vector (the direction in which light oscillates) matches the orientation of the photosensitive molecules, a signal is generated, and many such signals from across the microvilli add up. In this way, bee receptors can detect the plane of polarized light: if a receptor's "toothbrush" is fully aligned with the direction in which the wave swings, it will send a maximum signal to the brain. If the cell is pointing perpendicularly to the plane of the wave, a minimal signal will result.

This arrangement to detect polarized light has only been confirmed in the dorsal area of insect eyes—the area that looks at the sky during normal forward flight. In other areas of the bee eye—for example, the frontal or ventral area of the eye that might look at flowers—polarization sensitivity could cause confusion, because

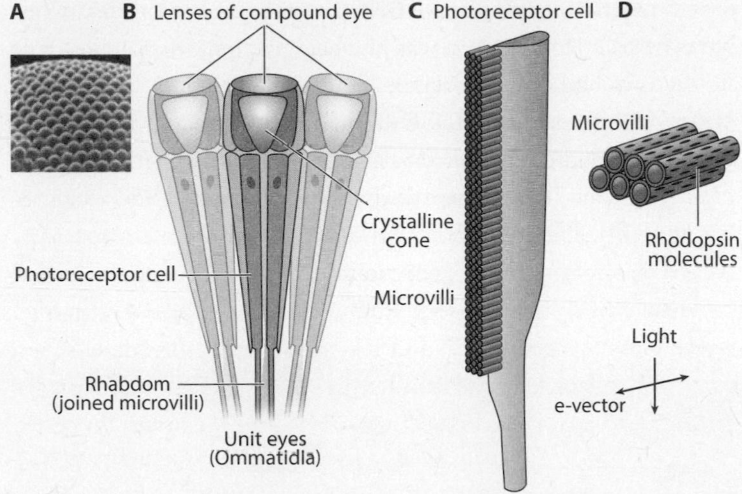


**Figure 3.3. Arrangement of structures mediating polarization sensitivity in an insect eye.** ***A.*** Curved surface of an insect eye with several dozen facet lenses. ***B.*** Three individual units—"ommatidia"—in longitudinal section. ***C.*** "Toothbrush" design of a single insect photoreceptor cell. The cell body is shown on the right (with the axon—its neuronal cable to the brain, indicated on the bottom). The arrows show the direction of light and the orientation of polarization. ***D.*** Photosensitive molecules (black bars) are oriented in a single direction; maximal stimulation occurs when the light reaching them is polarized in the same direction.

these very same receptors are also used for sensing color. Every biological receptor cell has only a single readout—the signal it sends to "higher up" in the brain. Now what if it had two input variables, so that its response varied with wavelength *and* polarization? Your brain couldn't easily make sense of that receptor signal; it would be like a gauge that measures both air temperature and pressure, but gives you only a single output on a scale from 0 to 100. Therefore, a useful receptor should only respond to one variable, and different receptor cells should ideally specialize in different functions (and not mix, for example, sensitivity to the plane of polarized light and the wavelength of light).

Accordingly, those bee photoreceptors that do not actually look at the sky—the frontal and ventral ones that look down at flowers, or ahead at landmarks such as mountains or trees—have a trick to eliminate their inbuilt polarization sensitivity. Their "toothbrushes" are twisted around their longitudinal axis. Thus, rather than all their microvilli pointing in the same direction, in these twisted receptors they all point in different directions. Biological nano-engineering par excellence!

## Sensitivity to Earth's Magnetic Field

As mentioned, Martin Lindauer later also discovered honey bees' sensitivity to Earth's magnetic field. Lindauer was able to manipulate the magnetic field around a beehive using Helmholtz coils—two large parallel magnetic coils placed on either side of the hive, which in combination can generate a near-uniform magnetic field, and can also be used to neutralize Earth's magnetic field. Lindauer found that in subtle ways, this affected the way bees communicated about food sources inside their colony. There is now evidence that they also use a magnetic compass outside their home—for example, to find the correct direction toward a food source. (A variety of invertebrate species share this superpower.)

When I worked in Stony Brook (near New York) as a postdoctoral fellow in the 1990s, I was allocated a basement laboratory space with no windows. While this was unglamorous, it resulted in a peculiar discovery that could not have been made in any conventional laboratory with access to natural daylight. Even though there was complete darkness in this laboratory once I turned the ceiling lights off, I often found that the bumble bees I was working with had been active overnight and had emptied a feeder more than a meter from their nest. Infrared video recordings revealed that the bees *walked* along trails like ants—a bizarre sight when I inspected these videos. I don't know if you have ever tried orienting in *complete* darkness (I mean not a dark night but absolute darkness, such as you'd find in a windowless cellar—the "dark dining" restaurants that have sprung up in many cities are a useful experience). You are completely orientation-less and helpless. You move at snail speed, feeling your way along the walls with your hands, and still bumping into things. But my bumble bees walked fast and were clearly oriented—and it turned out that, like ants, they used scent trails to mark the way.

But even when scent marks were completely removed, bees still oriented in the correct, familiar direction of the feeding station. It's likely that bumble bees used Earth's magnetic field for this task, too (though a definite proof awaits an experimental setup that manipulates the magnetic field). Two bits of advice from this: What you think are suboptimal working conditions (such as a lab space in which you never see the light of day) can sometimes lead to unexpected discoveries. And—the good old notion that science must be hypothesis-driven is fine for many purposes, but if you stick to it too strictly, it can limit your perspective: you can only discover things that are already on your radar. It's a good idea to keep your eyes open for completely unexpected phenomena.

The mechanisms, or sense organs, by which bees detect Earth's magnetic field remain controversial; there is some evidence that magnetoreception is achieved through iron granules located in the abdomen of bees. In one study, bees learned that the onset of a magnetic field reliably predicted a sugary reward (figure 3.4). When the researchers severed the nerve cord that connects the abdomen to the brain, bees failed to learn the association—indicating that

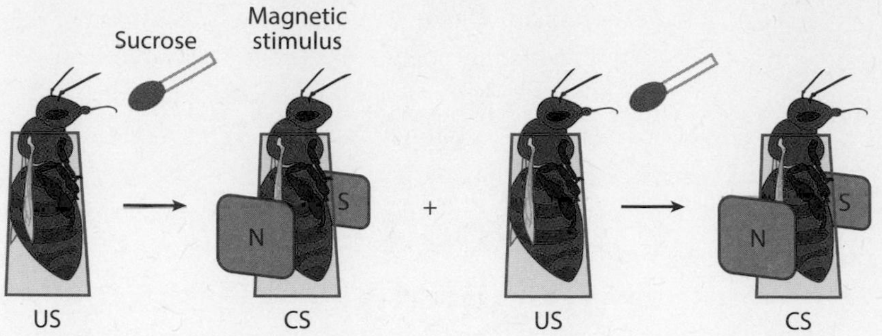


**Figure 3.4. Training bees to magnetic fields.** ***Left to right:*** When honey bees are offered sugar solution (US—unconditioned stimulus) on a cotton swab, they extend their tongue to drink. When exposed to a magnetic field (CS—conditioned stimulus), bees remain initially unresponsive, but when the CS is paired with the US several times, they learn the association and extend their tongue when exposed to the magnetic field, expecting that a reward is about to materialize.

the magnetoreceptors are indeed based in the abdomen, not in the brain itself or in the antennae. Such bees still managed to associate sucrose rewards with scents, indicating that the operation did not impair their learning abilities altogether.

## Antennae, the Strangest Sense Organs

Perhaps the most peculiar sense organs of a bee are its antennae (figure 3.5). Imagine you had two additional arms (but fingerless ones) attached to your head, but these arms have no power to lift things. Instead, their sole function is acquiring sensory information about your environment. They can smell, taste, hear, sense temperature, humidity, air currents, electric fields, and analyze

**Figure 3.5. A honey bee antenna.** Packed with sensory hairs and other microsensors, an antenna can smell, taste, and hear, as well as feel textures, measure temperatures, and respond to electric fields. Green: flagellum; blue: pedicel; inside the bend between the two (purple, plus adjoining zone of pedicel) is the bee's "ear," called Johnston's organ. The small basal part of the antenna with a socket is called the scape. Orange: eye; pink: surface of bee's head.

shapes and surface textures. Though all insects have antennae, such sensors are of course highly useful if you spend a good part of your time in the complete darkness of a bee nest. There, the antennae are constantly in motion (unless the bee is asleep), probing and assessing the multitude of sensory stimuli in the colony. When a bee lands on a flower, the antennae are used to locate the source of the desired reward, using multiple sensory cues. The antennae have three main segments, called *scape*, *pedicel*, and *flagellum*. The high mobility of the antennae is facilitated by a ball-and-socket joint between the scape and the head capsule (allowing rotational movements) and a further hinge joint between the pedicel and scape.

The entire surface of each antenna is tightly packed with sensors of a variety of types. The most conspicuous ones are hair-shaped, with various classes that differ in diameter and length—though these hairs are tiny (10–20 μm long). Many of them are used in olfaction, the sensing of airborne chemicals from a distance. Those hair-shaped protrusions that are used in olfaction (others are used in the touch sense, also called mechanoreception) contain multiple pores, underneath which sit the actual cells that are stimulated when an odor molecule binds to them. Other olfactory receptors sit underneath small pits in the surface of the antenna or under oval-shaped areas that are level with the surface of the antenna. In total, about 65,000 olfactory receptor cells are distributed along each worker honey bee antenna, of over 100 different types, each sensitive to different compounds. Mammals can have over 1,000 different types, but this does not mean that a bee is sensitive to fewer odors, for each type of bee odor receptor is sensitive to a different range of odorant molecules, and many odors and odor mixtures can be defined by a certain ratio in which the different receptor types are stimulated. A bee can even smell substances that we humans cannot, such as carbon dioxide—surely a useful asset in the crowded conditions of the beehive, where active ventilation can be crucial to survival if $CO_2$ levels are too high—which might also indicate that oxygen levels are critically low.

The high diversity of odor receptor types compared to bee color vision, which has just three types, makes sense when one considers the extreme diversity of airborne chemicals that matter in the lives of bees. A single flower species may produce dozens of odor molecules—and remember, there may be several dozen flower species in a bee's flight range. In addition, there are multiple pheromone signals that the bees themselves generate—larvae signaling that they are hungry, workers announcing danger or the discovery of food, queens asserting their dominance. Each of these pheromones is in turn a mixture of multiple different molecules. Inside the darkness of the nest, there are also many other odor cues (not generated by bees) whose detection is crucial—such as from mold on the comb structure, or from an intruder honey bee from a different nest who has come to steal honey.

Randolf Menzel discovered that honey bees can learn to associate odors with rewards extremely swiftly. For some odors, especially flower-like ones, a single reward paired with that odor can lead to 90 percent recognition accuracy, while for other odors that are typically less important in a bee's life, up to 10 trials may be required. Honey bees exhibit extreme flexibility with respect to such learning, so that they can associate sugary rewards even with scents that will never be encountered in natural flowers. For example, they can learn their own alarm pheromone—an airborne substance normally released by bees in the presence of threat, which induces aggression and stinging—as a predictor of reward. During late-night laboratory work as a young student, I once discovered that I had accidentally conditioned the bees to the scent of beer on my breath, so that they would extend their tongue in anticipation of reward when I exhaled.

The sensitivity of bees to odors is not as extraordinary as that of a sniffer dog—in fact, for several substances it is about on a par with humans. Nonetheless, several researchers experimented with the possibility of using honey bees as "sniffers"—for example, at airport security. The reason these attempts were ultimately unsuccessful is not because bees were incapable of learning, and subsequently responding to, the scent of explosives (they could detect them all right), but because of the frequency of certain types of errors. False alarms in airport security are fine (a human inspector can always subject a piece of baggage to closer examination), but you cannot tolerate "false negatives"—your sensor not responding when the signal is actually present. Bees might well make too many "false negative" errors to be suitable for use in airport security.

But the odor perception of insects may be superlative in the animal kingdom in another regard: its speed. Paul Szyszka, a student of Randolf Menzel, discovered not only that bees' odor receptors respond to an incoming odorant within 2 milliseconds; they can also detect a sequential appearance of two odors separated by just 6 milliseconds and distinguish this from a situation in which they are presented simultaneously. Thus, as bees move their antennae around their environment (for example, while inspecting a flower or a potential intruder to the hive), they may establish a fine-grained temporal profile (an "odor movie") of the inspected object that helps them identify it with high certainty.

## How Bees Taste with the Antennae

The senses of smell and taste are both forms of *chemoreception*; they are distinguished in that tasting is "contact chemoreception"—you have to touch the source of the chemical to taste it. Karl von Frisch explored the sense of taste in honey bees in detail. As in humans, there are many fewer types of taste receptors compared to the number of olfactory receptor types. Bees have taste receptors not just on their tongue and mouthparts (where one might expect them), but also on their feet (they can literally taste something they have accidentally stepped into) and on their antennae. As with other sensory modalities, what we perceive as similar may be distinct for the bee and vice versa. For example, our sweet receptors can be tricked with various artificial sweeteners—but bees cannot be tempted with saccharine. In addition to receptors for sweetness (obviously of importance in nectar foragers), bees also have a salt receptor, and they respond aversively to sour substances. Karl von Frisch found only moderate sensitivity to "bitter"

in honey bees, but bumble bees respond strongly aversely to some bitter substances, such as quinine. The responsible receptor has not been found in either species. Tragically, bees may respond positively to some (possibly bitter-tasting) neurotoxins found in pesticides applied by farmers to agricultural crops, such as neonicotinoids, which also find their way into floral nectar.

## Feeling and Hearing with the Feelers

Other hairs on the antenna are *mechanosensors*. They are the tactile (touch) receptors that give the antenna its other name—a "feeler." Some of these hairs measure the position of the various antennal segments relative to each other (i.e., proprioception—the sense of one's own body postures and movements), whereas others are used for sensing external stimuli. They are hollow protrusions of the *cuticula* (the chitinous exoskeleton of insects), often anchored in such a way that they bend predominantly in one direction when pressure is applied. As the bee is constantly palpating objects of interest (such as a flower or, in honey bees, a wax comb under construction, but also moving objects inside the hive, such as other bees or potential intruders), the integration of the bee's voluntary antennal movement with incoming mechanoreceptor information can thus give information about the shape and identity of an object. Outside the nest and when visiting flowers, the antennae are also used to feel the fine structure of flower surfaces, helping the bee to find the path to the reward.

But not all of a bee's mechanosensors come in hairlike structures. Inside the link between the antenna's distal or outermost part (the flagellum) and the middle part (the pedicel) is Johnston's organ—a set of mechanoreceptors that measure how much the flagellum is bent relative to the pedicel (figure 3.5). It turns out that these receptors deliver a bee's sense of hearing. People have long taken a dim view of hearing in bees, dating back to John Lubbock's early observation of the bees' indifference to his violin playing, and due also to the observation that bees lack the type of eardrum that we have (as do many insects with a need for long-distance hearing, such as crickets—but they have them in their forelegs). The so-called tympanic membrane in such eardrums is structured to measure pressure changes generated by sound waves. But it turns out that the Johnston's organs in bees' and other insects' antennae measure a completely different aspect of sound: they gauge the actual air particle movements that make the flagellum vibrate, rather than pressure waves.

Using this peculiar mechanism, honey bees can hear aural communication signals emitted by other bees, albeit only over a few millimeters' distance, and only over a range of about 20 to 500 hertz (cycles per second). For comparison, young humans' hearing has a frequency range of 20 to 20,000 hertz. Many of the sounds that *we* hear bees make, for example when we listen to the rich variety of sounds in a beehive, are not perceived by other bees as sound at all, but as vibrations of the honeycomb, sensed by the bees' legs. In the darkness of the cavities in which some species of bees nest, where bees seeking information about food locations can't see a dancer, both substrate vibration and airborne sounds are useful for detecting the position of a dancing bee.

## Bees' Sensitivity to Electric Fields

Sigmund Exner (1846–1926), Karl von Frisch's uncle and also an important scientific mentor of his, had already discovered that bird feathers can accumulate electric charges by friction with air. However, it took until 1974 for scholars to explore the potential biological significance of this phenomenon in bees. In that year, the Russian scientists Evgeny Eskov and Alexander Sapozhnikov found that honey bees can carry large electrostatic charges—in fact, any flying object, whether it's an insect, a football, or a jumbo jet, loses electrons and is therefore positively charged. Charges can also be generated by rubbing body parts against each other. The team not only discovered that bees are surrounded by electric fields, but that bees' antennae can sense such fields. They do this not by any specialized electric field sensors, but by antennal mechanoreceptors subjected to Coulomb forces—the mechanical attraction or repulsion that results from two objects carrying opposite or like charges. Note that you can sense electric fields without specialized receptors, too—for instance, when an electrostatically charged balloon makes your fine hairs stand up. The researchers suggested that these fields might play an important role in bee dance communication (see chapters 5 and 8) inside the hive. In honey bees, the most important mechanoreceptors for electric fields appear to be in the same antennal organ that detects sound: Johnston's organ was found to respond to such electrostatic charges.

A British team found that bumble bees also respond cleverly to the electric fields of flowers. Flying bees are positively charged, whereas flowers are literally grounded—and thus negatively charged. When a bee visits a flower, the electric charge transfer results in the flower becoming temporarily more positively charged. This temporary "electrical imprint" on the flower tells other bees that the flower has been recently visited, and is therefore not worth visiting again, since nectar takes some time to replenish after a bee has drained a flower. In addition, flowers display characteristic electrostatic patterns that can be sensed by bees as they explore the flowers and which they can use as "invisible nectar guides" to locate the rewards efficiently. This team found that the mechanosensory hairs that cover a bumble bee's body might respond to such charges: as a positively charged bee moves over a negatively charged flower, the hairs might bend in a particular, predictable manner.

Thus both in honey bees and bumble bees, devices normally used in sensing mechanosensory stimuli also mediate the sensitivity for electric fields. How can one disentangle two types of sensory stimulus—vibrations/touch and electric fields—with the same sensor? The answer is that bees probably can't—unless there are auxiliary sensors to indicate which type of stimulus is currently most salient. And perhaps they don't have to. It is quite possible that, to a bee, the two types of stimuli are felt as variations of the same modality.

Imagine humanity was suddenly cast into a world of darkness, where the curious hair-raising experience that occurs with electrostatically charged balloons took on biological relevance. Suppose that every time you feel the sensation, this precedes bumping into an obstacle. Not only would you swiftly learn to use this sensation—you might learn to use any hairy part of your body to explore your environment to avoid collisions. Yet you could still distinguish the electric charge–induced mechanosensory stimuli from others—say, someone stroking your hair. And unless your physics teacher told you, you might not even be aware that one source of stimulation is physical contact, while the other is electrostatic. Over many generations, such hypothetical humans might even evolve particularly high sensitivity to electrostatic forces. So long as there's an environmental stimulus that has biological relevance, animals are sure to make use of it (if they can sense it at all, with whatever means)—either through individual experience, or over many generations through evolutionary processes.

We humans have our senses of sight, hearing, smell, and taste neatly packaged into completely different sense organs (and we can't additionally taste with our feet). This is clearly different in bees, whose antennae are as multifunctional as a Swiss Army knife! Our fingertips are at least *somewhat* multifunctional: we can feel the texture and shape of an object, as well as measure its temperature and humidity. But imagine you could also smell it, taste it, hear it, and measure its electrical charge at the same time with the same fingertips. The sensory world of insects is strange, and rich.

In this chapter we have learned how any information that enters the bee mind from the external world is first sieved through the sensory filters that they have acquired during evolution. But information in the mind is not only obtained in each individual's lifetime. Our instincts, honed over many millions of years, inform us about potentially hazardous situations, and possibly palatable and unpalatable food; they furnish us with basic forms of locomotion and determine when we respond to others with aggression or affection—and much more. In the following chapter, we will discover the rich repertoire of instinctual behaviors in bees. Such instincts not only determine, in large part, what is on any animal's mind, but also what that animal can learn.

# 4

# "It's Just Instinct"—or Is It?

These observations show us how flexible is the instinct of the bees, how well it yields to the local conditions, to the circumstances and requirements of the family. Necessity, in the work of these insects, as in all that pertains to the habits of animals, must be limited to a small number of essential points, all others being subordinated to the circumstances. . . . The limits of their industry are assuredly less narrow than at first supposed; and the reader will admit, with us, that the conduct of bees depends also in some measure upon what might be called the judgment of the insect; this judgment, doubtless, is rather a matter of tact than formal reasoning; but its subtleness resembles choice, rather than habit or habitual mechanism independent of the will of the insect.

**—François Huber, 1814**

One might wonder why a book about the mind—any animal's mind—needs a chapter on instinct. Instincts strike many of us as atavistic drives that a mind is to rise above and control—that are simple and automatic, in contrast with a mind's freedom and complexity. In fact, many biology students will tell you that they think the main difference between human and nonhuman animal behavior is that nonhumans do everything by instinct, and humans nothing.

But, of course, what is on our minds *is* governed by instinctual drivers: to mate, to care for children and relatives, to feed, to fight for survival when threatened, and so on. You don't need to learn individually that large animals with big teeth are probably a threat, or that feces aren't yummy food (it's different if you're a dung beetle). Nonetheless, evolution has left us relatively few rigid guidelines for how to cope with these challenges. In humans and in many other animals, even the most elemental behavior routines need to be refined by learning: instinct provides little more than a rough template. You need to learn to suckle on a nipple, to walk, to fight, to detect a predator, to defend yourself, even to have sex. The *solutions* that humans have developed to cater to our instincts for behaviors such as sexual gratification, resource acquisition, or attack and defense against competing tribes are impressive. Yet it is useful to remind ourselves that most of our behavior, and in fact a lot of what is on our minds, is still governed by elementary drivers related to survival and biological fitness. We use our intelligence to cater to needs governed by our instincts.

We will learn in this chapter that, in bees just as in us humans, instincts are much more than the toolkit of a robotic animal. They are interfaced with learning at multiple levels, and almost invariably allow for considerable behavioral flexibility. They also determine what an animal *can* learn, and the limits of its learning. Humans, for example, have a "language instinct" that distinguishes us from all other animals: we are preprogrammed to learn to communicate by language. But the details—the vocabulary, the language-specific grammar—need to be learned.

We will begin this chapter by learning that the instinctual behavioral routines of bees can be highly sophisticated. Even behaviors that appear governed by innate routines, such as wax comb building, are rarely entirely preprogrammed, and not only need to be partially learned, but involve high degrees of flexibility and even, possibly, planning skills. The diversity of instinct-governed behaviors found in the hymenopteran insects (which include not only bees but also wasps and ants) is possibly without parallel in the vertebrate world. Among these insects, there are some whose instincts determine them to be neurosurgeons (wasps that paralyze their insect prey with exactly three venom injections—one to each nerve center in the thorax), farmers (ants that tend aphids like cattle, or grow fungi for food inside their nests), perfume collectors (males of certain bees that douse themselves with volatiles to impress females), harvesters of solar energy (flower visitors that collect the sugars generated by plants' photosynthesis), and builders of mathematically perfect multistory structures (the honey bees' wax comb; figures 1.1, 4.2).

There is no question that an insect of a certain species has no say over whether it will be a flower visitor or a hunter; this is indeed determined by instinct. But these instinctual dispositions fill the mind: they determine what is *experienced*, as rewarding or as a threat; and they determine what is *thought about*. Animals don't just adapt a stereotyped repertoire of solutions to the challenges of their species-specific lifestyle; they sometimes invent wholly new solutions by searching their mind-space. In this chapter, we will learn that, in bees just as in humans, instincts interface seamlessly with memory and cognition. Instinctive predispositions promote learning ability, and intelligent behavior can emerge from evolved innate behavior.

## Jean-Henri Fabre and the Notion of Insects as Reflex Machines

An influential champion of the idea that insects are essentially inflexible in their elaborate, but innately programmed behavior was the French entomologist and writer Jean-Henri Fabre (1823–1915). Fabre came from a peasant background and struggled with poverty for most of his life; a teacher in provincial schools, he was an autodidact in his scientific endeavors as an entomologist. With minimal means and no university affiliation, he produced an oeuvre of several dozen scientific books, among them the ten famous volumes of the *Souvenirs Entomologiques*, an incredible treasure chest of scientific discovery. His detailed, insightful observations of, and clever experiments on, insect behavior made him the founding father of the discipline of animal behavior; his engaging writings earned him two nominations for the Nobel Prize in Literature.

But Fabre believed in neither Darwin's theory of evolution nor the intelligence of insects. And some of his experiments didn't exactly present insects' wits in the most favorable light. Here's an example in which insects displayed what Fabre called "their machine-like obstinacy."

Pine processionary caterpillars form long, single-file "head-to-tail" queues. Each individual produces a silk and pheromone trail that is followed by other caterpillars. In one experiment, Fabre tricked a train of caterpillars up onto the rim of a large vase (135 cm in circumference). They walked around the edge, laying a scent trail, and the leading caterpillar soon found its own trail . . . and the procession continued in a circle . . . and continued . . . and continued—a total of 335 times over the course of seven days, always in the same direction. Concludes Fabre: "These figures surprise me, though I am already familiar with the abysmal stupidity of insects . . . whenever the least accident occurs. I feel inclined to ask myself whether the Processionaries were not kept up there so long by the difficulties and dangers of the descent rather than by the lack of any gleam of intelligence in their benighted minds." To be fair, the caterpillars did manage the descent of their own accord, on the eighth day, following exploratory activities of a single descending scout on day six, after Fabre had placed some tasty pine needles near the bottom of the vase. But this clearly isn't a glorious case of quick insect problem solving.

Fabre isn't much more generous to the hymenopterans, but he tries. "Where do you find a more richly gifted animal? . . . Can the bird, this wonderful architect, compare his works to the construction of the honey bee, this masterpiece of higher geometry? Even man has a rival in the Hymenopteran. We build cities, so does he; we keep servants, so does he; we breed domestic animals . . . he has his milk cows, the aphids," he muses, and then goes on to say, "To consider the animal means to ask the disquieting question: Who are we? Where do we come from? Thus: what goes on in this tiny Hymenopteran brain? Are their abilities related to ours, is there a form of thinking? . . . What a Chapter of psychology, if we could write it!" (Fabre fittingly calls this chapter of the *Souvenirs Entomologiques* "*Fragments* of the Psychology of Insects.")

So Fabre explores whether such hymenopteran innovations might, at some level, be the result of intelligence; his conclusions are fairly damning. He makes various manipulations of the nest constructions of a mason bee, a solitary bee that builds a single clay pot for each egg / (later larva)—the pots look a bit like tiny swallow nests, only the bee seals the opening hole once provisioning is complete. In one experiment, Fabre drills a hole into the base of the brood chambers under construction, and observes that the bee will simply complete the construction at the top, even intermittently adding more nectar to the pot while the honey is leaking from its base. The bee appears unable to depart from its predetermined path of construction, and as gifted as she is in masonry skills, she is unable to fix the obvious damage that will condemn her larva to death. Fabre's verdict: "Oh little ray of reason, of which they say that you illuminate the animal—you are very close to darkness, you are nothing."

Only once does the famous French entomologist depart from this stance. Fabre observed the hunting habits of various species of digger wasps. Many of these wasp species dig burrows in which to raise their larvae. Unlike those of bees (which descended from such wasps in evolutionary history), their larvae are strictly carnivorous, and their moms provision them well: depending on the wasp species, they will provide their young with caterpillars, adult insects of various kinds, or spiders to feed on. In contrast to social bees, which directly feed their larvae, however, the wasp mother faces the challenge that she will be long gone, and possibly dead, when the baby wasp, a tiny and helpless grub, hatches from its egg. Thus, the mother needs to ensure that the food will still be palatable several weeks after she herself is gone.

Fabre wondered how the wasp's prey was kept so fresh for several weeks, when he knew well that any dead insect would decay within days in the summer heat. He dissected the insects that the wasps had deposited in their burrows—and discovered that they were not just "fresh" but actually alive: they were paralyzed, and the wasps' larvae were slowly eating them alive. Fabre observed that the wasps were highly skilled at immobilizing their prey without impairing their vital functions. The three pairs of legs (as well as the one or two pairs of wings) of insects are controlled by one nerve center (ganglion) each in the thorax. Knock out these centers, and only these, and the prey is immobilized but can still breathe, albeit slowly.

Fabre discovered that digger wasps of the genus *Sphex*, which hunt insects with this structure of the nervous system (three ganglia in the thorax), give their prey exactly three venom injections—one to each ganglion. In some prey species, however, such as some beetles, the three ganglia are fused into a single nerve center. Wasp species that hunt these apply only a single injection into that center, whereas wasps that hunt large caterpillars inject every ganglion of the body, to ensure that their prey is immobilized head to tail. Through evolutionary trial and error over many generations (mutation and selection of the most efficient methods), these wasp species have specialized in the anatomy of their particular prey's nervous system.

One species, which Fabre calls the yellow-winged Sphex (presumably *Sphex flavipennis*), hunts crickets. Like other species of digger wasps, it constructs a burrow for its young first, before embarking on a hunting excursion. If successful, it brings the paralyzed prey back to the nest, often over considerable distances. Invariably, before the victim is placed into the burrow, the wasp briefly abandons it next to the entrance, and disappears into the cave it has dug, perhaps to ensure that no parasites hide inside or that it hasn't partially collapsed. If everything is in order, the wasp then reemerges and pulls its unfortunate quarry into cricket hell (figure 4.1).

Fabre wished to explore how fixed this series of actions is, and so, while the wasp was busy inspecting the cradle for its young, he moved the cricket a few inches away from the nest entrance hole. The wasp reemerged, appeared surprised to find its prey so far away, and then dragged it back to the hole's entrance—where it again deposited it to inspect the hole. Fabre repeated this about forty times, and it never occurred to the wasp not to leave the cricket unguarded, but to pull it straight in. The philosopher Daniel Dennett uses this as an example of the "mindless mechanicity"

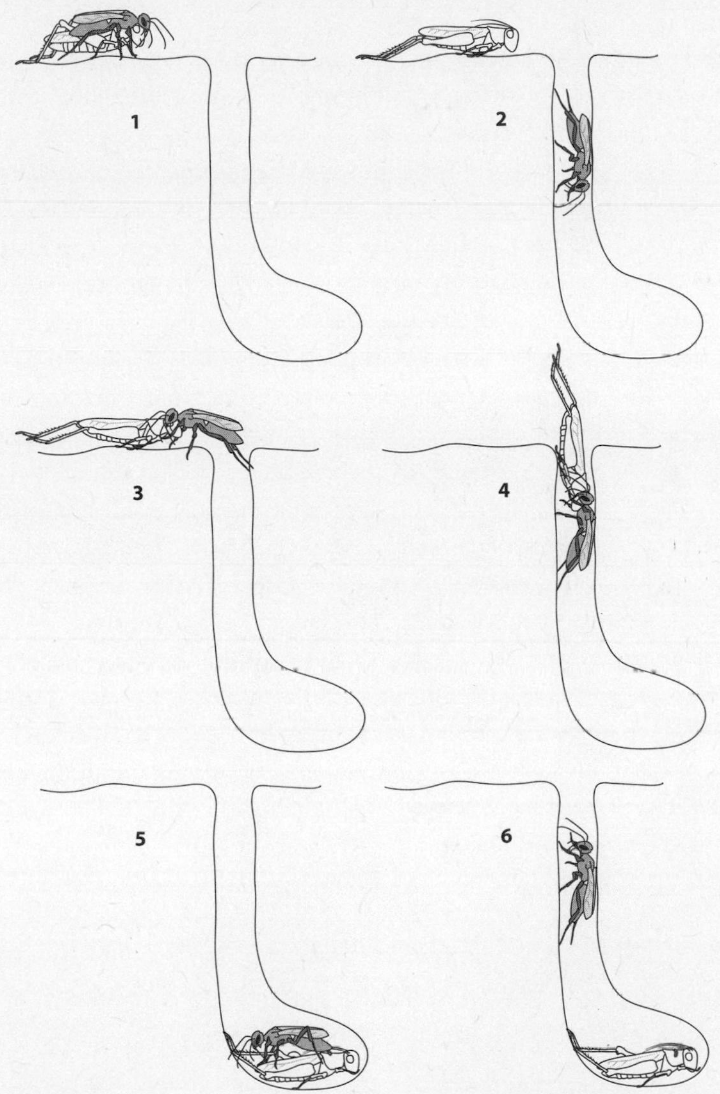


**Figure 4.1. Sequence of actions of digger wasp depositing a paralyzed cricket into its burrow.** ***1.*** Arrival and deposition of prey next to the entrance. ***2.*** Inspection of burrow. ***3–4.*** Dragging of prey into burrow. ***5.*** Laying of egg on prey. ***6.*** Departure from nest.

of insect behavior, and gives it its own term—"sphexishness"—to contrast it with the flexibility of human behavior and the free will that supposedly governs it.

Maybe Dennett didn't read the paragraphs that followed Fabre's description of these wasps' seemingly stereotyped behavior. Fabre reports how he tried to play the same trick on individuals of another population of the same species of digger wasp. He writes: "after two-three times . . . the wasp . . . grabs its prey by the feelers and drags it down. Who was the fool now? The experimentalist, who had been outwitted by the clever Hymenopteran! . . . Intelligence is heritable . . . there are clever and less clever kinds; and this apparently depends on the abilities of the ancestors."

How remarkable that Fabre should have expressed these views! The entire *Souvenirs Entomologiques* is filled with profound admiration for the sophistication of these insects' instinctual behaviors—and with equal contempt for Darwinism and for those who credit insects with intelligence. In all ten volumes of Fabre's masterpiece, there is just this one episode in which he appears to have briefly changed his mind about the "cleverness" of insects, and, in passing, affirmed two key ingredients of Darwin's evolutionary thinking: variation between individuals, and the fact that such differences are heritable.

## How Instinct and Intelligence Combine in Honey Bee Comb Construction

The view is still widespread that hardwired routines can explain the entire diversity of social insect behaviors. But when people have looked closely, it has become evident that even behaviors long believed to be entirely genetically preprogrammed need to be partially learned—and are in fact remarkably plastic. Let us examine the case of honey bee wax comb construction, the *Apis* solution to housing larvae as well as storing food (figures 1.1, 4.2).

Darwin noted that the "comb-making power of the hive-bee" was "ranked by naturalists as the most wonderful of all known instincts." It is, at first sight, a wonder of animal architecture. The dexterity it requires for a six-legged animal to manufacture a repetitive structure with such regularity and precision is remarkable. The resulting construction is also a case of highly optimized engineering: hexagonal cells are better than the round cells of bumble bees, since the latter arrangement wastes a lot of space between cells. Square or triangular cells would have no gaps between cells, but since the larvae to be raised in the cells aren't square or triangular, you'd waste lots of space inside the cells. So hexagonal cells are a good idea, and in fact some species of wasps build them too, albeit of paper rather than wax.

But no other genus of bee except honey bees also builds double-sided hexagonal combs, which is yet another impressive trick to save space. The bottom of each hexagonal cell has the shape of a pyramid (again a more efficient solution than a square bottom), and the two sides of the comb interface perfectly with one another through these pyramid-shaped bases of the cells. This double-sided structure wouldn't be a good idea for the horizontal combs of stingless bees—gravity being what it is, you couldn't easily store honey in a vessel whose orifice points down. So honey bees build their comb structure vertically, with each cell opening sideways. But if cells were aligned perfectly perpendicular to gravity, this still wouldn't work too well for liquid contents (as you may quickly discover if you hold an open jar of liquid honey on its side). You need to keep it at an angle so that the honey's viscosity and adhesion to the vessel keep it inside. And that's precisely what honey bees do: the cells of the comb are tilted slightly downward from the opening to the base. Multiple combs are built in parallel, leaving just enough space for workers to move about freely. This is despite the highly irregular shape of the cavities in which some species of honey bees nest naturally, such as hollow trees (unlike the cuboid boxes beekeepers supply them with).

So far, so ingenious, but just instinct, right? At first glance, the repetitive structure of the comb seems like the perfect result of some robotic, hardwired behavioral routine—an assembly-line job of building the same structure over and over. But is it really this simple?

The most detailed and insightful observations of wax comb construction were recorded by the Swiss naturalist François Huber (1750–1831). Blind from an early age, he was assisted in his investigations (and presumably in writing them up) by his wife, Marie-Aimée Lullin (1751–1822), and his servant, François Burnens (1760–1837). Using glass-walled observation hives, Huber, Lullin, and Burnens laid the foundations for the scientific exploration of honey bee biology. The level of detail they recorded and their experimental exploration of wax comb construction in the bee colony were not just unprecedented then—they have also not been rivaled since.

Their book describes how many variations there are in the comb structure; for example, the first row of cells differs from subsequent ones, since it functions as a foundation. One might suspect that worker honey bees use their own body as a sort of template to arrive at the correct dimensions of each comb cell—but this is certainly only part of the story, since the cells destined for drones are 30 percent wider (and they are also built by workers). There are multiple other modifications of wax structure, such as for the wholly differently shaped cradles for queens, or for pillars and crossbeams as support structures to stabilize combs. Huber's team observed in detail how comb construction is initiated by a single worker at the top of the hive, and how multiple individual workers sequentially contribute to the construction of each cell. Different workers continue cells where others have left off (and do so correctly no matter the previous state of the cell), and inspect one another's constructions to amend them where necessary. On one occasion, it was observed how a worker misplaced a piece of wax, and another one swiftly corrected the error.

Huber and his team deliberately explored the bees' flexibility in comb construction. They began by testing bees under conditions in which they could not attach honeycomb to the ceiling of the hive, as they will normally do. In that case, the bees built their comb from the bottom up, essentially a tower construction rather than a hanging one—reversing many of the motor routines that they would normally use for building the cells top to bottom. Next, Huber et al. prevented the bees from building either up *or* down. In that case, they started at one of the side walls and extended the comb laterally across the cavity.

But here comes the real chef d'oeuvre: while the bees were busy extending their comb laterally through the hive (eventually to connect it to the opposite wooden wall), the experimenters covered the target wall with glass, a suboptimal surface to attach the comb to. Huber anticipated that perhaps once the bees had reached the glass, they would make some sort of special efforts to attach the comb to this slippery surface. But they did something else altogether: apparently noticing that their intended target surface had been rendered suboptimal, the bees turned the construction of their comb by 90 degrees—*before* it reached the target wall (figure 4.2). Huber reported that he repeated this experiment in multiple ways, sometimes moving the glass target into the projected path of their comb building several times, and bees would change the direction of their construction again and again. He observed that bees had to change the dimensions of

**Figure 4.2. An experiment to probe the flexibility of the honey bees in comb construction in the face of unusual challenges.** Swiss entomologist François Huber (1814) had noticed that bees avoid, when possible, attaching the comb construction to glass walls of observation hives. ***A.*** When bees were faced with the hive that had a glass ceiling and floor, they would begin their construction on one of the side walls. ***B.*** Before bees had reached the target wall, a glass screen was placed over that wall. Rather than continuing the construction in the same direction, bees introduced a curve into the construction. Continued construction of the comb in the revised direction resulted in adhesion to a more suitable target area for attachment.

the hexagonal wax cells around the resultant kink in the comb—the cells were two to three times wider on the outside surface than on the inside. He mused about how so many bees "agree" on changing the direction of the construction, and the truth is that we still do not know.

To further expand on the remarkable plasticity displayed in comb building, Huber reported how honey bees cope when disaster affects their construction. To enable monitoring the bee colonies' inner workings over extended periods, Huber experimented with glass hives. If nothing else is available, bees *will* use glass to attach combs to, but, as we will shortly see, this is not straightforward. During winter, foraging for flowers and brood rearing are halted, and bees will minimize any activity to ensure their stores last until spring. On one occasion in winter, one of several combs broke off the ceiling in one of Huber's glass hives. Not only did bees wake up to fortify the dislodged comb with a number of pillars and crossbeams made from wax, but they subsequently also reinforced the attachment zones of all the other combs on the glass ceiling, to ensure that a similar calamity wouldn't happen again. Says Huber: "I may restrain myself from reflections and commentaries, but I acknowledge that I could not suppress a sentiment of admiration for an action in which the brightest foresight was displayed."

One might counter that the precautionary repairs induced by the breakage in the beehive need not necessarily be based on foresight—that instead they might be yet another hardwired routine triggered by a certain stimulus configuration. Perhaps. But you should also consider whether relegating such preventive behaviors to yet another innate building routine is any "simpler" an explanation than claiming that they do require a form of planning.

In arguing for or against a role for cognition vs. prewired "instinct," you can't rely on intuitive impressions of which path to the same behavior *looks* simpler or more complicated. For a scenario in which *all* of the different construction behaviors reported by Huber are governed by instinct, you'd have to consider how many ready-made neural circuits in bees' brains it would take to mediate all of them. You'd also have to explain how they could all have evolved even if some of the tested challenges—such as the sudden appearance of a glass pane in the path of comb construction—will never have been encountered in evolutionary history. Could it be that a cognitive scenario—in which bees have a multipurpose mechanism for *understanding* the outcomes of their actions—is actually a simpler explanation? Perhaps bees do have a sort of mental "master plan" for the desired outcome of their building activities, as Huber suggested.

Whichever way you look at it, bees do not behave fully in line with what you'd expect if comb construction were governed wholly by instincts. In natural bee comb constructions, there are a variety of subtly different ways in which wax comb is structured. What's more, the way in which young workers build comb is affected by the structure of the comb they were raised in and had the opportunity to sample for some time after emergence. Naïve individuals raised in circular plastic cells, unassisted by experienced workers, will manage to build hexagonal honeycomb—but the structure of the comb is far from regular, and the cell diameters are all over the place. As with many other behaviors that are commonly regarded as strictly instinctual, such as spider web construction, innate predispositions may only provide a rough template for baseline behavior. The details often need to be learned, can be adapted flexibly to environmental conditions, and may be subject to planning.

A hive of bees once traveled on board the Space Shuttle *Challenger*, two years before its doomed final mission in 1986. The honey bees spent an entire week in zero gravity. Not only did they learn to fly under such conditions, but they built honeycomb with cells of normal dimensions. The only difference (compared with honeycomb built on Earth) was that the cells of space honeycomb were not consistently angled downward—not surprising, since there is no obvious "down" in zero gravity. But the geometry of the combs was correct: several combs had the usual straight, flat structure, and were built roughly in parallel, in the complete absence of gravity.

## Simple, but Erroneous, Early Explanations of Bee Behavior: the "Homing Sense"

In interpreting possible instances of animal intelligence, it is always useful to look for ways to explain the behavior without invoking intelligent problem solving. But sometimes scientists have been so dogmatic in their search for "simple" explanations that they were blind to clear manifestations of learning behavior. An example is Albrecht Bethe (1872–1954), a German physiologist, the father of nuclear physicist and Nobel laureate Hans Bethe (one of the developers of the atom bomb). Albrecht Bethe knew from others' work that bees and ants can find their way home from distant places with extreme accuracy. He also thought that learning was too complex an explanation. So he tried very hard to find an instinctual force that drew bees back to their hives. To disrupt this force, he tried spinning the bees hundreds of times, gluing magnets to their backs, and more. Nothing worked—the bees still found their way back home, unless they were carried too far away. Bethe's conclusion: "They are not guided back to their nest through memory images, or acoustic, or magnetic, or chemical stimuli . . . there is nothing left but to assume that bees are guided home to their hive through a wholly unknown power." He preferred to invoke an enigmatic "unknown power" rather than concede that insects might actually memorize the location of their nests.

The logic of how Bethe reached this conclusion is even more revealing. He noticed that when he displaced a beehive by just a few meters, the returning bees would search at the previous location of the nest entrance, rather than fly directly to their hive box, which was plainly visible from the old location. He concluded that they cannot have used memory—because he thought the only possible way to solve the task was to memorize the appearance of the hive itself, rather than the landmarks surrounding it.

In the title of a magnificently argued reply to Bethe's work, the German zoologist Hugo von Buttel-Reepen (1860–1933) asked (in 1900!): "Are bees reflex machines?" According to Betteridge's law of headlines, any title that ends in a question mark can be answered by the word "no," and Buttel-Reepen's paper is no exception. He highlights Bethe's interpretation of the beehive-moving experiment as anthropocentric, since Bethe clearly overlooked the possibility that there is more than one way to solve the task. From a bee's perspective, the response that Bethe expected—to recognize and enter a home that has been moved a few meters—is suboptimal at best and fatal at worst. In many species of bees, both solitary and social, multiple nests can be in close proximity to each other, so it is of paramount importance to enter only a home that is in the exact correct location. Any error might result in feeding the larvae of an unrelated mother—or being killed by guard bees.

Buttel-Reepen also presented plenty of sound evidence (including from Bethe's own work) that bees knew the correct location of the hive from nearby landmarks that they had memorized. The lesson is that in deciding what is the simpler explanation for intelligent-looking behavior, one has to avoid the pitfall of explanations that *seem* simple, but are simply wrong. Assuming that the Earth is flat sounds simple at first, but flat-earthers have to come up with very convoluted narratives to explain away the facts.

## Are Bees Attracted to Flowers "by Instinct"?

Buttel-Reepen further dissected the question of whether bees can be considered "reflex machines" in the context of flower visitation. To this day, in many publications on pollination ecology and flower evolution, you find the notion of "flower attractiveness." Implicit in this term is the idea that flowers, with their showy colors and pretty scents, are simple trigger stimuli that appeal to the airborne "reflex machines" that populate the skies, and somehow cause them to drop onto the flowers, without regard for the flowers' reward levels or any information from previous experience. Clearly, this idea was already outdated in 1900: Buttel-Reepen points to the evidence that bees will return to a location where they have previously been fed even if their feeder has been removed—and thus there is no stimulus to trigger a reflex. The only information that guided the bees to the location came from their memory.

Likewise, Buttel-Reepen pointed out that bees do not "automatically" land on a flower when they come across its signal: certain flowers present nectar only in the morning, and he noticed that bees will ignore the stimuli of the flowers at times of day when bees know them to be unrewarding. Nonetheless, even today the idea of "automatic," reflex-like responses of pollinators to flowers is still implicit in the concept of "pollination syndromes"—the idea that pollinator classes (such as bees or beetles or flies) have narrow affinities to certain flower attributes ("hummingbirds visit only red flowers, nocturnal moths only white flowers"; figure 4.3).

**Figure 4.3. The pollination syndrome concept holds that different pollinator types have narrow affinities to certain types of flowers.** In this view, hummingbirds are tightly linked to red flowers, bumble bees to blue flowers, and nocturnal moths to white flowers (continuous arrows). However, since most pollinators learn about rewards rather than simply being attracted to certain flower features, there are many interactions (dashed lines) that defy the stereotype.

Much as the Bohr model in physics is perhaps a useful simplification to introduce atomic structure to schoolchildren, the concept of pollination syndrome is still taught to young ecologists at university as a dogma. Again, beware of simple stories.

In the early 1990s, just after the Berlin Wall had collapsed, I discovered a lovely nature reserve to the east of Berlin, the Naturschutzgebiet Lange Dammwiesen near Strausberg. There, I attempted to monitor as many interactions between flowers and pollinator classes as possible over three sequential field seasons. Since a substantial part of my duties as a PhD student was teaching, I took armies of students to this field site to observe flower patches for months on end and note down all the visitors to the flowers. The students were not instructed on what they might be expected to find, so the sampling was as unbiased as it could be. This was the first attempt at quantifying an entire pollinator-plant interaction network. The results, published in an article with pollination biologists Nick Waser, Mary Price, Neal Williams, and Jeff Ollerton, showed no statistically significant difference in the flower color affinities of bees, flies, butterflies, and beetles. This is as one might expect if pollinator choices are largely determined by individual learning rather than innate preference.

## Learning and Instinct Evolve Hand in Hand

In the evolutionary history of bees, a key event was the innovation of provisioning their young in a specially constructed nest, which required not just (instinctual) construction skills, but also a precise spatial memory. Add to this the (again instinctual) lifestyle of harvesting floral nectar and pollen. This involves learning about, and making economic choices between, multiple flower species that differ in the quality and quantity of rewards they offer, and the signals that advertise these rewards. Again, the instinct that determines the lifestyle facilitates and indeed necessitates learning, and if the innate predispositions of some individuals allow them to learn faster, fitness benefits will accrue.

The ability to learn about new environmental contingencies can also facilitate the evolution of innate preference. For example, suppose there is a new, nectar-rich species of flower that smells of turpentine. (Remember that bees are sufficiently flexible even to associate their own alarm pheromone to a reward, so learning "implausible" scents is not impossible.) Among local pollinators, only individuals with the flexibility to depart from "conventional" flower scents and swiftly learn the association between turpentine and reward will be able to exploit this new resource, and they will have an advantage over individuals that visit only other flower species. Within the population of rapid, flexible learners, there subsequently arises the additional possibility of evolving, over many generations, an innate preference for turpentine that enables locating the flowers even more efficiently. Some studies have shown that changes in scent preference can come about by adjusting the same synapses in the nervous system either over evolutionary time or by individual experience. If ability X is "evolvable" in a small brain as an instinct, there is no reason to assume that one can't also "innovate" it in a small individual brain, sometimes even by making similar neural circuitry adjustments.

There are *some* species of pollinators, including some species of bees, that appear to have relatively narrow "inborn" affinities to certain flower species or other food sources. The advantages and risks are obvious: if you can identify your food from information you have inherited rather than having to acquire it by individual exploration, you can save yourself a lot of sampling time and effort. On the other hand, if you are narrowly specialized and your favorite food happens not to be available, you're in trouble. And so, upon close inspection the purported affinities for certain foods are often less strong or exclusive than was once supposed. For example, many flower visitors that were once reported to visit only certain flower species "by instinct" can also be found exploiting other flower species when their typical food is scarce.

Most social bees are generalists, but a bumble bee species called *Bombus consobrinus* reportedly visits only the highly complex but also highly rewarding flowers of monkshood (or devil's helmet) in parts of its range. As a specialist, this bumble bee has an advantage in efficiency over generalist visitors exploiting the same flowers. But even *Bombus consobrinus* has to learn to handle the flowers—though even entirely naïve individuals have a head start over any member of other, generalist bumble bee species. Once again, it's obvious how innate predisposition and learning interface. A generalist may be armed with only a general notion of what constitutes a rewarding flower, and have to figure out the rest from scratch, by trial-and-error learning. A specialist is born with a basic "operating manual" to its preferred food source, acquired over many generations, and just has to fill in the details by individual learning.

There are thus interactions between innate behavior and cognition at multiple levels. For example, honey bees are faster than most other animals at learning to associate colors with rewards—not because bees are more intelligent than, say, cats, but because colors have much less meaning in the life of a cat than they do for an animal that obtains all its nutrition from flowers. Instinct begets cognition, and even phenomena that were largely thought to be governed by innate behavior, such as wax comb building, are neither simple nor explainable without learning and cognition.

It remains an open question why so many unusual instinctual behaviors have emerged in the Hymenoptera and are much rarer in other insect orders and beyond, but it is clear that bees are not simple reflex machines. One key instinct that has especially promoted the remarkable learning abilities of bees in their evolutionary history is their need to memorize their spatial environment. We will learn in the next two chapters that as central place foragers—animals that have a nest to whose location they must unfailingly return—bees must be highly accurate in their spatial memory.

# 5

# The Roots of Bee Intelligence and Communication

How did the honeybee obtain its mental faculties? The comprehension of the gradual evolution of mental abilities in an animal species as a natural process is a task of such extraordinary interest, that no effort, however tiring, must exasperate us. . . . The perfection of this family branch occurred . . . with the transition to society formation and the concurrent division of labor . . . among the native bees, in bumble bees and honey bees.

**—Hermann Müller, 1876**

The famous Austrian ethologist and Nobel laureate Konrad Lorenz thought that the challenges our forest-dwelling primate ancestors faced operating in a complex three-dimensional environment were crucial to the evolution of human intelligence. He proposed that the mental exploration of spatial possibilities—for instance, when deciding whether a jump to a nearby tree is feasible or potentially fatal—has the key ingredients of all thought processes, including the most advanced abilities in humans, such as language.

That's eminently plausible, but humans' ancestors are not the only animals that have to perform complex operations in three-dimensional space. Indeed, we have learned in the last chapter that bees face significant spatial challenges, not just in their natural habitat, but also in the construction of their own elaborate three-dimensional nesting architectures, and that this may involve certain forms of planning. In this chapter, I will explore the possibility that spatial learning may be at the root of the evolution of bees' cognitive abilities, and of a symbolic communication system about spatial information that is unique among all animals: the honey bee dance "language."

## The Triassic Ancestors of Bees—the Cruelest Carnivores

To explore the early roots of the evolution of the mind of the bee, we have to go way back in time. The Hymenoptera (the order of insects containing the bees, ants, and wasps) first appeared in the Triassic, about 220 million years ago (figure 2.5). Dinosaurs roamed the planet, and there were most likely no flowers yet. Like many insects, such as flies or butterflies, the ancestral hymenopterans were solitary, and they were vagabonds—they did not have a home to provide shelter for their young, nor did they provision them with food. Females deposited their eggs on vegetation, and the larvae fed on plants. We know this from many of their extant descendants, which continue to follow this lifestyle.

In the early Jurassic, however, an important lifestyle change occurred: some Hymenoptera switched from a vegetarian diet to a special form of carnivorous lifestyle. Rather than depositing their eggs on vegetation, they began laying eggs on live animals, often herbivorous ones that sat on vegetation or burrowed inside it. Larvae would hatch from these eggs and consume their host alive. This lifestyle is called *parasitoidy*, as distinct from *parasitism*, where the death of the host is not necessarily part of the plan. From the moment a parasitoid wasp places its eggs on you, you have almost no chance—the outcome is already all but certain.

Since animal hosts typically pack protein more densely than plants, this lifestyle change would have been advantageous for the newly minted carnivore—but only given the sensory capabilities and nervous system to locate its mobile (and often hidden) prey. This is a considerably bigger challenge than just dropping eggs on a leaf, because animals can defend themselves, and hide. Many extant parasitoids are extremely adept at locating hosts—such as insect larvae hidden under the bark of trees or burrowed deep inside a fruit—and then inserting their ovipositor precisely in the right spot to lay their eggs on the doomed larva.

Many species are proficient at learning the chemosensory and visual cues that indicate the presence of a suitable host, and some employ spatial learning, both to avoid laying eggs on the same host more than once and to relocate to places where finding hosts is more likely. Indeed, some of the most remarkable feats of spatial learning occur in parasitoid wasps. For example, females of the scorpion wasp *Hyposoter horticola* monitor the sites of suitable butterfly host eggs in their environment. The only period for the wasp to lay its eggs on those of the host is immediately before the host larvae hatch, so upon discovery of the target the wasp does not deposit its own eggs right away, but instead remembers the location of the host eggs, returns to them intermittently over weeks to monitor their progress, and then returns for oviposition exactly at the right moment. This is a simple, and perhaps largely hardwired, form of prospecting for future opportunities, and of planning.

Using the same logic we developed in chapter 2 about the evolution of color vision, it is possible to reconstruct the brain structure of the ancient Jurassic Hymenoptera (which include the ancestor of the bees). That is, by evaluating the brains of extant animals, and using the simple logic that biological traits are generally more likely to stay the same than to change, we can extrapolate back to the animals' ancestors. Using such an analysis, Sarah Farris and Susanne Schulmeister of West Virginia University discovered that, concomitant with the Jurassic lifestyle switch from a herbivorous to a parasitoid lifestyle in a branch of the Hymenoptera, a significant change appeared in brain structure. The "mushroom bodies" (see chapter 9), higher-order brain centers in insects, which function in multisensory integration, learning, and memory, are dramatically enlarged (and more "convoluted," like a mammalian cortex) in the parasitoid Hymenoptera (and their bee descendants), compared with their herbivorous ancestors. Clearly, the new lifestyle required additional computational equipment.

There are of course significant risks of leaving your eggs on a mobile host such as a beetle grub: the eggs might become dislodged, or the host (along with the parasitoid eggs or larvae) might fall victim to a predator, such as a woodpecker (or, back then, perhaps a small dinosaur). And so, sometime around the beginning of the Cretaceous (~140 million years ago), another important transition happened in those parasitoids that would later give rise to the bees. The wasps started excavating nests which would provide a shelter for their eggs and growing larvae. To these nests they would bring live prey to make sure their larvae had sufficient food—and they paralyzed the prey, to ensure that it could not shake off the parasitoid's egg and at the same time to warrant that the food was kept fresh for weeks (as Jean-Henri Fabre discovered, recounted in chapter 4).

The evolutionary innovation of such central place foraging required spatial memory as a prerequisite, and it placed new demands on the precision of that memory. The prospecting scorpion wasp that forgets the location of a suitable butterfly egg patch can find another one. But evolution won't be forgiving of a mother who forgets the location of her own home that contains her offspring. With this lifestyle, errors are not tolerated.

Jean-Henri Fabre was the first to investigate spatial learning in insects, and he highlighted the many challenges faced by the digger wasps of the genus *Ammophila* (which can tend up to three separate nests simultaneously, and up to ten over a summer). Fabre was unsurprised that honey bees can find their hive, as it is large, smells of bee, and has constant flight activity at its entrance. But the digger wasp seals her burrow with pebbles and sand to remove any trace of its existence (figure 5.1). She can only recover its location from memory, which is not easy, since there are often multiple other nests of conspecifics in the near vicinity. Fabre emphasized that the learning evinced by digger wasps must be one-trial learning—they often dig their burrow in the evening, then spend the night elsewhere, and return the next morning without fail (their nest is for the offspring only, not for themselves).

It is likely that the brain transformation that coincided with the earlier switch from a vegan to a carnivorous lifestyle came in handy when an even more precise spatial memory was required for central place foraging. (Having a nest, in turn, would—much later—facilitate the evolution of sociality.) Curiously, one branch of the parasitoid wasps reverted to a strictly vegan lifestyle and became the flower-visiting bees (perhaps 120 million years ago; figure 2.5). It is important to know that many adult parasitoid

**Figure 5.1. A solitary digger wasp sealing her nest entrance with a pebble.** Bees evolved from parasitoid wasps, of which a representative is shown here. Digger wasps of the genus *Ammophila* sometimes maintain multiple nests at the same time; they close their nesting burrows with pebbles and sand to protect their offspring and food, and sometimes even use pebbles to pound and compact the sand that seals the entrance. Locating these hidden nests among the multiple neighboring nests built by conspecifics places considerable demands on the wasps' spatial memory.

wasps also visit flowers for nectar—and may well have done so back then—and could have accidentally brought home some pollen in the process. Some must eventually have swapped meat for floral pollen (also extremely rich in protein) as larval protein provision, and stuck with it ever since. Hermann Müller (1829–1883; see introductory quote), was a German biologist who sought to apply Darwinian principles to understanding the evolution of plant-pollinator interactions. He suggested that in addition to larval provisioning and nest building, a further ingredient in the evolution of bee intelligence was the need to handle ever more complex flower morphologies.

## Karl von Frisch and the Discovery of the Bee Dance Language

In the preceding paragraphs, we have learned that a major transition in brain evolution in the hymenopteran insects preceded the evolution of sociality. The key behavioral innovations that were accompanied by the elaboration of the mushroom bodies were the evolution of nest building (and the required spatial memory) as well as the need to detect and discriminate between a wide variety of food sources. These challenges are still shared among the many thousands of solitary bee species that exist today, and the few hundred that are social. The advent of sociality, however, necessitated and facilitated the evolution of further behavioral capacities—most notably communication. Many forms of information sharing have evolved in social bees, but we focus here specifically on one, the honey bee "dance language." We do so because this symbolic communication system provides a unique window into an insect mind, allowing us to explore how bees perceive the spatial environment about which they communicate, and also because there has been considerable work on its evolution and adaptation to the environment.

In the final years of World War II, while Karl von Frisch was forced by historical events to work on applied honey bee research, he returned to a peculiar phenomenon he had already observed twenty years earlier. Von Frisch knew that successful scouts who had returned from a rich floral food source would display a rather strange behavior on the vertical wax comb inside the hive. In what looked like a "dance," such foragers would engage in highly stereotyped, repetitive motor displays, sometimes for minutes, and such "dancers" were eagerly followed by other bees inside the hive. Already in the 1920s, von Frisch had surmised that these dances had a function in communication about floral food resources. He had observed that there was variation in the pattern of the dance, and (incorrectly) conjectured that some types of dance indicated that the scout had discovered a valuable pollen source, whereas others alerted foragers to a useful nectar source. In 1945, after von Frisch had moved his research from the bombed-out university building in Munich to rural Austria, he had several eureka moments. In the months surrounding the end of the war, von Frisch discovered that the bees' dances contained codes for the direction and distance of the food source the scouts had located. A symbolic communication system about spatial coordinates in an insect—the nearly sixty-year-old biologist had finally made the discovery that was to win him the Nobel Prize.

Briefly, this is how the dance language works. The successful forager wags her body from side to side, moving forward in a straight line (the waggle run). Then she runs in a half circle to the left, back to her starting point, performs another straight waggle run along the path of her first, and then circles to the right (figure 5.2). This pattern is repeated multiple times, and is eagerly attended by unemployed bees in the hive. Shortly after such dances commence, dozens of newly recruited foragers arrive at the food source being advertised.

Von Frisch found that the angle of the waggle run from the vertical is equal to the angle between the sun's azimuth (its compass direction) and the flight direction of the food source from the hive. For example, if a food source is found in the compass direction of the sun,

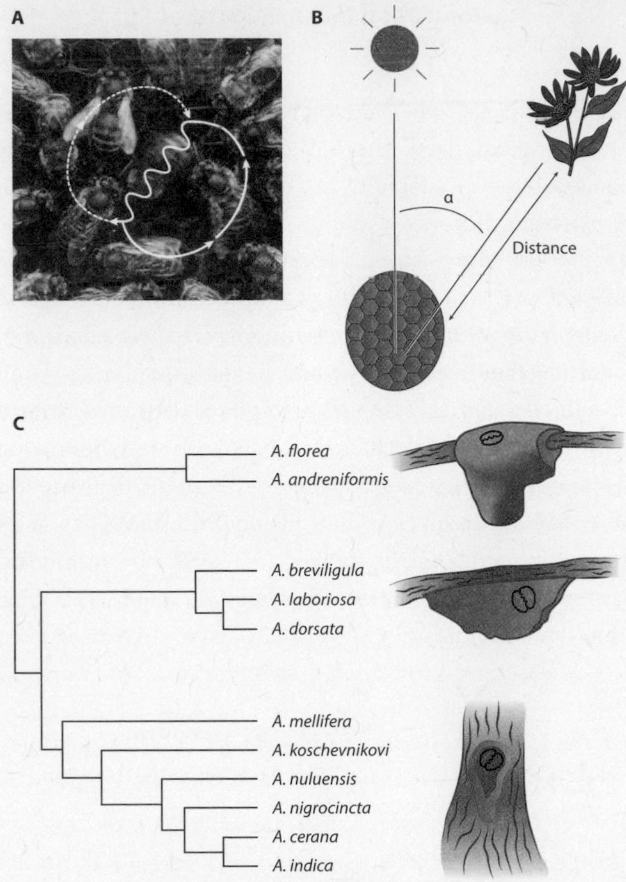


**Figure 5.2. Figure-eight-shaped waggle dance of the Western honey bee (*Apis mellifera*) and its Asian relatives**. A waggle run oriented 45° to the right of "up" on the vertical comb (***A***) indicates a food source 45° to the right of the direction of the sun outside the home (***B***). The abdomen of the dancer appears blurred because of the rapid motion from side to side. ***C***. The dance might have evolved from a species whose comb (similar to the top two extant species in the phylogenetic tree) was attached to a tree branch, such that dances are performed on the near-horizontal top surface of the wax construction. In these species, bees do not use gravity as a reference, but instead their waggle run points directly to the destination. In the middle cluster of three species, open combs are attached to thick tree limbs or rock projections on cliffs, but dances take place on a vertical comb surface with gravity as a reference. This might have been a useful preadaptation for cavity nesting (e.g., in hollow trees) as occurs in the bottom cluster of six species, where gravity *must* be used as a reference, since no celestial cues are available as compass cues inside the dark nest.

the dancer will waggle straight up the vertical comb. If food is found 45 degrees to the right of the sun's direction, the waggle run will be oriented 45 degrees to the right of vertical on the comb (figure 5.2). The distance to the target, a flower patch with abundant nectar or pollen, is encoded in the duration of the waggle run: the longer the bee waggles, the larger the distance of the food from the hive.

Remember, it's typically dark inside a beehive, so other bees cannot visually observe the dance but must decipher it by following and sensing the dancing bee as she moves through multiple dance circuits. The dance followers must learn the indicated location information while attending to the dance, decode the information, and then apply it at a later time in a very different space from where it was acquired. No other species (besides humans) uses a similarly symbolic representation to communicate about spatial locations in the real world.

## The Evolution of the Dance Language

How could such an extraordinary communication system have evolved? In a half-year expedition from 1954 to 1955, Karl von Frisch's assistant Martin Lindauer traveled to Ceylon (now Sri Lanka) to study the closest relatives of the domesticated western honey bee: several other species of the genus *Apis*. Through comparisons between related, extant species, Lindauer hoped to decipher a proto-dance language—the earliest evolutionary roots of the honey bees' unique communication system. Among the species he explored were the fierce *Apis dorsata*—hornet-size, highly aggressive bees that build giant, single combs in the open, suspended under overhanging cliffs or from tree branches—and the dwarf honey bee, *Apis florea*, another open-air nester also nesting in trees.

Lindauer discovered that all honey bee species displayed the characteristic figure-eight run (albeit with slightly different "dialects" for the direction and distance codes). In all species, direction to a food source is assessed relative to the current azimuth position of the sun. In all species except the ones considered closest to the ancient ancestor of all of today's honey bee species (*Apis florea* and the black dwarf honey bee, *A. andreniformis*), this angle is expressed during the dance (which is performed on a vertical surface) as the angle relative to gravity. *A. florea* and *A. andreniformis*, which nest in the open, and whose dancers dance on a horizontal surface, do not perform the transformation relative to gravity: their foragers orient their dances relative to the direction of the sun, as it is seen on the flight to the known flower patch. Lindauer thought this to be the original form of the dance.

This is a plausible scenario, since it suggests that the dance evolved in an open-nesting bee that did not yet have to make the transposition to gravity. But the origins of the tight interaction that occurs between dancers and recruits in the hive still remained enigmatic. How did dancers and recruits first establish contact? This may have been facilitated by the possibility, in open-nesting bees, of recruits *seeing* the dancer. Nonetheless, the question remains: which evolved first, the peculiar behavior of successful foragers to display a highly stereotypic motor pattern, or the readiness of potential recruits to follow successful foragers? Clearly, both need to exist for recruitment to work. But why would any bees follow a successful forager if it had not yet "invented" a message that, at the very least, identified it as a successful forager? And why would successful foragers perform excited motor patterns in the nest if no one had yet evolved the predisposition to attend and follow?

Martin Lindauer subsequently traveled to the tropics again, this time to South America, where he teamed up with the Brazilian biologist Warwick Kerr (1922–2018) to find out if the stingless bees, then thought to be the sister group of the honey bees, might provide information about the earlier roots of the honey bees' dance language. Unfortunately, they did not provide the desired answer. None of their numerous species seem to perform the kinds of highly repetitive motor patterns called dances in the honey bee. Many, however, do perform excited runs inside their nest once they return from a successful foraging bout, possibly to arouse inactive foragers to engage in foraging. Several species also emit small vibrational pulses during these runs, using their thoracic flight muscles. In some species, these pulses are correlated in length with the distance of a food source, as are the waggle runs of honey bees.

Besides these shared traits, stingless bees show strong differences in recruitment systems between species. Some use scent trails; others appear to guide nestmates directly to a food source. Some stingless bees appear to make "intention movements"—repeated short flights from the nest ("false starts") in the direction of the food and back—which other bees could interpret as a source of information for the direction of the food. It is possible that similar intention movements were performed by an early, open-nesting ancestor of the honey bee, and that this formed the basis for the direction code in the waggle dance.

## An Ancient Version of the Dance in Bumble Bees?

Decades later, in the late 1990s, my then master's student Anna Dornhaus (now a professor at the University of Arizona, Tucson) explored the communication of the bumble bees, now known to be the actual evolutionary sister group of the stingless bees and, up till then, thought to lack any form of communication system about food sources. Anna discovered that the bumble bees in fact have a highly efficient recruitment system—a single successful forager can alert the entire forager force of the colony by running around the nest in irregular patterns (and spreading an alarm pheromone). But in bumble bees, this communication lacks any spatial information whatsoever. Recruits have to find the flowers themselves, though they do obtain from the successful foragers the scent of the rewarding flowers.

In conclusion, the common ancestor of the social bees (honey bees, stingless bees, and bumble bees) likely did perform excited runs inside the nest after finding food. But these took separate evolutionary paths in the three groups, evolving into a large diversity of communication systems; none retain a "behavioral fossil" version of an ancestral honey bee dance. How the ancestor of honey bees evolved, step by step, the abstract, symbolic communication form of the dance language, we will probably never know. There is simply too large a gap between the behavior of the honey bees and that of their closest relatives.

## Why Do Honey Bees Dance?

Another way to ask the question of how the dance language might have evolved is to measure its fitness benefits under natural foraging conditions. There are no mutant honey bees that don't dance, however, so we had to resort to a trick to generate "mutant" honey bees in which the information content of the dance language was scrambled. As part of her PhD thesis work, Anna Dornhaus tilted the combs of honey bee hives into a horizontal position, so that bees could no longer use gravity as a reference. Under such conditions, *Apis mellifera* will use the sun (or an artificial light source, if an experimenter installs one) as a reference, presumably as its open-nesting tropical ancestors did. But if only diffuse light is made available, the dances become disoriented, so that waggle runs will point in random directions even within a single dance, from one waggle run to the next. Recruits still follow such dances, but they receive no direction information. They might still receive distance information, which will, however, be relatively useless without knowing which direction to fly in.

At a field station near the University of Würzburg, Anna then compared the foraging success of two groups of hives, those with disoriented dances and those whose communication was undisrupted. The result: there was no difference. This was a bit of a surprise—the honey bee dance is widely heralded as one of the most spectacular examples of animal communication, indeed a Nobel Prize–winning discovery. Yet taking it away from the bees had no effect on their foraging performance.

Perhaps the setting in which we performed these experiments, an agricultural landscape in Bavaria, did not present floral resources in a spatial distribution that mimicked the natural conditions under which the honey bee dance evolved. Anna decided to repeat the experiment, this time in Spain's Sierra de Espadán nature reserve, with more than 300 square kilometers of largely unspoiled nature. And again—disrupting the information flow between dancers and recruits had no appreciable effect on colony foraging success.

We began writing up this rather stunning result for publication, but then decided that we should repeat the experiment one final time in a different location. All honey bee species except the western hive bee reside in tropical Asia, and all of them have the dance language. It is therefore straightforward to assume that this communication system evolved in a tropical setting. Floral food in tropical forests is very differently distributed than in temperate habitats. In the latter, flowers are often distributed widely in space—think of any temperate natural flower meadow, where bee food is dispersed, not aggregated. In tropical forests, however, much of the floral food comes from trees, and an individual tree might provide thousands of flowers—but the nearest tree in bloom might easily be a kilometer away. In between, there is green, green, and green—it looks lush to us and perhaps to herbivores, but to a flower-seeking bee, it's a desert. In a tropical forest, floral food is thus often extremely aggregated in space.

Anna decided to set up her experiment once more, in Bandipur National Park, once the Maharaja of Mysore's private hunting reserve in South India. The first challenge was to assess the spatial distribution of floral food to see whether indeed it might be more clustered than in temperate habitat. If you've ever seen a tropical forest, this task is plainly impossible for a single person over an area that's anywhere close to a honey bee hive's foraging range (with a radius that can be over 10 kilometers).

But could one not simply spy on the bees, by translating their very own dance language to see which food sources they find worth advertising? Using beehives as a "radar" for floral food, Anna discovered that, indeed, the locations advertised in the dances in a tropical setting were much more clumped than those in temperate habitats. And indeed, under such conditions interfering with the dance communication system made a huge difference—it reduced the number of successful foraging days to one-seventh of the success rate of hives with fully functional communication.

It was the spatial clustering of floral food in the tropical forest, then, that generated the evolutionary pressure on honey bees to come up with a precise communication system about spatial coordinates. In temperate habitats, which were colonized only by a subset of honey bee species, the dance communication might be largely an evolutionary hangover from their tropical ancestry. It is, however, also conceivable that the dance language evolved in a completely different behavioral context—the need for bee swarms to locate a new home (on this, more in chapter 8)—and that once the communication system was in place for that context, it was then co-opted for conversing about flower locations.

We have learned that a key factor in the evolution of bee intelligence was the early evolution of central place foraging: the establishment of a home that contains one's offspring, and thus the requirement of a precise spatial memory. The ancestral (and still solitary) bees shared with their parasitoid forebears the need for a precise spatial memory of their nests. The fact that spatial memory in insects is now commonly studied in social species such as ants, honey bees, and bumble bees does not mean that such memory is any less important in solitary species. In the following chapter, we will explore just how complex a bee's spatial memories are, and how on some occasions researchers have eavesdropped on the honey bee dance communication to understand an insect's mental representation of space.

# 6

# Learning about Space

There is not the faintest doubt that bees . . . are finding their way home guided by the images taken of their hive and in its near and far vicinity. You can call it an instinct only insofar as instinct guides the bees on their first flight to inspect the location of their home and its immediate surroundings. . . . On their first flight, they turn around and fly first in small and then ever larger loops to obtain a precise picture of their hive.

**—Johann Dzierzon, 1900**

Jean-Henri Fabre experimented on the homing of multiple species of solitary wasps and bees. He caught them at their nests, marked them with colored paint dots, and carried them in containers (from which they could not see the way) to distances up to four kilometers in various directions. Invariably, a high fraction of the released individuals returned to their nests, sometimes only on the next day, but typically much more swiftly.

When Charles Darwin read about these experiments, he wrote to Fabre, suggesting various tricks to confuse the bees during displacement to find out the strategies by which bees locate their home. Fabre experimented with carrying bees in the opposite direction to the release site, spinning the bees around rapidly, carrying them on detours, releasing them behind hills—but invariably, most of the bees made their way home without fail, in spite of all these treatments. Upon reading this, Darwin suspected that bees might use a magnetic compass and suggested disorienting them by gluing magnetic needles onto their backs. Fabre duly tried to follow this suggestion, but the first bee managed to remove the needle after a struggle that is described in colorful terms, and Fabre never repeated the experiment.

In the end, Darwin and he (like Albrecht Bethe later—see chapter 4) agreed that pigeons and bees must have a special, unidentified homing sense that we humans lack. Curiously, neither of them considered what today seems the most obvious interpretation: that bees have a memory of the landscape around their nest, enabling them to fly home from any location inside their familiar home range. As will become clear in subsequent paragraphs, bees establish a rich library of spatial memories of their flight environment. Whether these are joined together as a sort of mental representation—a cognitive map—remains the subject of ongoing controversy.

## Bees Use Landmarks to Navigate

The first empirical demonstration that bees remember the location of home by using landmarks was provided by the African American scientist Charles Turner (1867–1923), a pioneer of research into insect cognition whose findings we encounter in almost every chapter of this book. Born just two years after the end of slavery in the United States, Turner began publishing groundbreaking work at the age of twenty-five, and published over 70 scientific articles in his relatively brief career. This was despite the fact that because of his ethnicity, he was denied a professorship at a research-focused university, and instead spent much of his life as a teacher at a school for African American children in St. Louis—without access to a laboratory, a library, or support from a team of researchers. Turner published widely on comparative brain anatomy in birds and invertebrates, individual variation in behavior and learning competences, and intelligent problem solving in a large variety of animals, at a time when dominant thought credited animals with only the simplest of learning abilities. Unfortunately, Turner's discoveries and conceptual advances failed to gain the recognition they deserved, and his works were later all but forgotten.

Turner tested the bee "homing sense" with an elegant and simple experiment in 1908. He observed a solitary burrowing bee whose nest entrance was close to a discarded Coca-Cola bottle cap. When the cap was moved to a nearby location next to an artificial burrow that Turner had made, the bee crawled into that burrow without hesitation—indicating that the insect had a memory for landmarks rather than, for example, being guided by an instinct to follow the scent of the nest. In several later studies, Turner confirmed his assertion that bees and wasps navigate by what he called "memory pictures." The mysterious "homing sense" of the hymenopteran insects that had so baffled nineteenth-century scholars such as Fabre, Darwin, and Albrecht Bethe (see chapter 4) was, at least in part, based on a memory for the visual scene surrounding the nest.

## Learning in Context

In the 1980s, scientists began suspecting that bees might have a richer representation than one that simply consists of memories for landmarks, panoramas, and flight vectors (and the ability to communicate about the latter). In one elegant experiment, one of the foremost scholars in animal orientation, the British biologist Thomas Collett, teamed up with then-junior Almut Kelber (now herself a leader in the field of animal vision) to study context learning in the spatial memory of bees.

They trained honey bees to visit feeding stations in two identical huts, placed 75 m from the hive and 33 m apart. The interior of the huts was identical—in both huts, the bees found an arrangement of two blue and two yellow cylinders placed in a rectangle (figure 6.1). In one hut, food was offered halfway between the two yellow landmarks, and in the other, halfway between the two blue ones. When the bees' search behavior was evaluated in the absence of any food, they correctly searched between the yellow or the blue cylinders depending on which hut they were in—even though the entire panoramas inside the huts at the time of decision making were identical. The bees must have retrieved the appropriate memory from contextual cues—in this case, either the landscape seen prior to entering the hut, or their own flight behavior (remembering which direction they had taken to reach the hut).

Such context learning has since been confirmed in many different experimental paradigms with honey bees and bumble bees. For example, bumble bees can learn to visit either yellow or blue artificial flowers depending on whether these are presented under blue or green illumination. A key feature of the Collett and Kelber experiment was that the contextual cue was only available from memory—not directly accessible in the real time of the choice situation. The bees thus had to use one memory (landmarks seen or actions performed before arriving at the hut) to access another (whether to land between the familiar yellow or blue landmarks inside the hut).

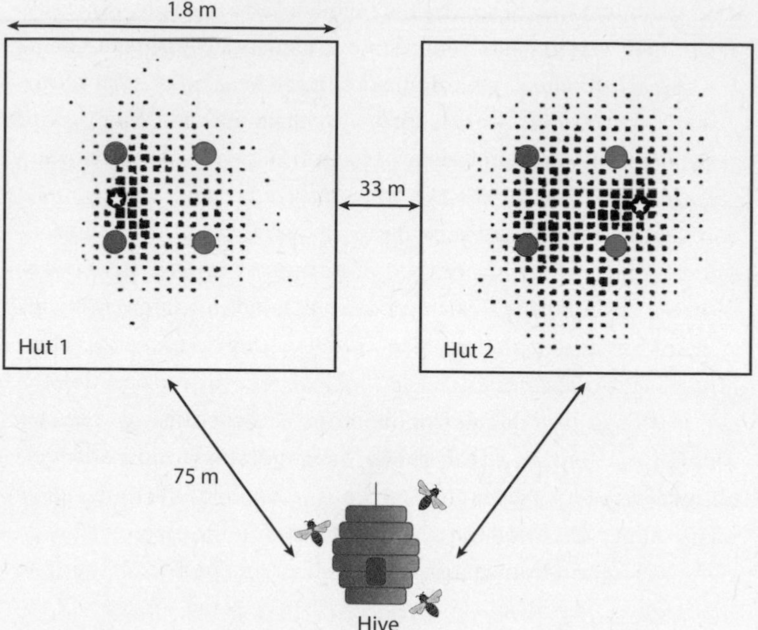


**Figure 6.1. Context learning in honey bees**. Honey bees learned to visit two identical huts with four cylindrical landmarks inside (figure not to scale). In one hut, reward was provided between the two blue cylinders, and in the other between the two yellow ones (reward locations indicated by asterisks). When reward was removed during tests, bees nonetheless searched predominantly in the correct areas (indicated by density and size of black dots), even though no cues inside each hut indicated its identity; bees thus decided based on recent memories from before entering the huts.

## Cognitive Maps in Bees?

Around the same time as the experiment about bees' context learning, the Princeton biologist James Gould suggested an even more advanced representation of spatial memory in the bee mind. He proposed that bees might have a *cognitive map*—a mental representation of their familiar environment, which they can essentially "imagine" and which allows them to perform spatial operations that would not be possible with simple route-based memories. Martin Lindauer's work contained anecdotal evidence that bees might have rather flexible access to their spatial memory library: he observed that some bees with known experience of a spatial location would dance at night, indicating that precise location, which must have meant correctly extrapolating the (invisible) sun's position behind the horizon at that time. More importantly, though, these observations indicate that bees can spontaneously retrieve a spatial memory in the absence of any external stimulation—a far cry from the conventional view that memories are simply triggered by incoming stimuli ("bee remembers the reward of the yellow flower when it sees one") or internal triggers ("bee's honey stomach is full, hence it needs to search for memorized landmarks pointing the way home").

But Gould went even further in terms of how flexibly bees might use their mental representation of space: he suggested that dance followers can essentially "look up" the advertised coordinates on their own mental maps, and assess whether they are plausible. He proposed, for example, that when dancers returned from a feeding station on a boat in a lake, the dance followers might inspect their internal maps, decide that the indicated location couldn't possibly bear reward, and might accordingly ignore the dances. As beautiful a hypothesis as this is, a later controlled experiment did not confirm it.

Gould also published a study indicating that bees might be able to take novel, previously untraveled shortcuts between familiar destinations, and that it might be their mental map that allows them to do so, rather than, for example, simply spotting the correct destination from a distance (figure 6.2). These were, at the time, revolutionary claims. Most insect learning was viewed as simply associative, like bees learning a flower color as a predictor of

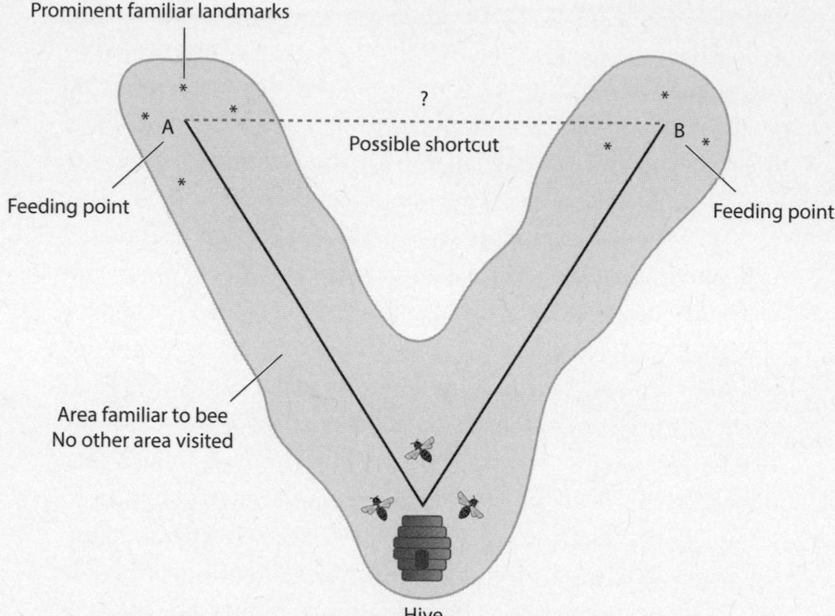


**Figure 6.2. Cognitive maps in bees?** Having a cognitive map allows animals not just to remember landmarks or routes, but also to compute novel spatial solutions based on those memories, such as novel shortcuts between familiar locations. For example, assume that a bee had previous experience with flying from the hive to A and from the hive to B. Using a cognitive map should allow this bee to travel a novel shortcut from A to B, whereas simple route memories would require the bee to fly via the hive to get from A to B.

reward. The idea that an insect might actually have flexible access to internal representations of the world around it, enabling it to compute novel spatial solutions, was seen by many as improbable.

The skeptics included Randolf Menzel, then supervisor of my master's project. In the summer of 1989, he took a whole crew of master's and PhD students from Berlin to the house of his in-laws in a small town in West Germany to put Jim Gould's ideas to the test. We had a small observation hive, and about 2,000 worker honey bees were marked with individual number tags. The plan was simple. We were going to first train the bees to fly to feeding station A, about 500 m from the hive. Then, over a period of two days, we moved the feeder along an arc with the hive at the center, several meters at a time, to a second destination B.

The question at the end of the training was simply this: Would bees whose known destination was B, which we caught upon their departure from the hive, be able to fly a shortcut to B when we displaced them in a dark container to A and released them there? Since the bees were labeled, we knew all the individuals that had visited positions A and B and all the stations in between. We reasoned that during the training procedure, these bees would have come to know the entire area, but would only have explored it in a fan-shaped pattern with the hive as the origin. They should not have had experience with flying directly from A to B, but a cognitive map might allow them to do so. However, upon release, the bees departed exactly in the same direction they would have taken had they not been displaced from the hive. If they had a cognitive map, they did not make use of it in this setting. This and several similar experiments from around the world seemed to settle the matter of the cognitive map in bees.

This study was to be the first scientific paper on which my name appeared as an author, and I remember feeling slightly uncomfortable with it, for several reasons. One was that a negative result in an animal cognition test never means that the animal is actually incapable of solving the task. A typical error from human navigation illustrates this. In my local supermarket, the vegetable shelf was at some point moved from the far back of the store to right by the entrance. I marched right past the desired vegetables to the end of my familiar path to find that the produce was no longer there. I had simply traveled a familiar vector to the end before even paying attention to my surroundings. Such errors do not mean that I am actually incapable of seeing a familiar sight at an unfamiliar location. In the same vein, bees may have initially traveled in the correct intended direction, and only have looked at landmarks near the end of the memorized travel distance. Another complication, however, was that the landscape used in these tests, like those used by others in similar experiments across the globe, was extremely cluttered. In such a setting, it was simply impossible to decide which local features the bees actually made use of.

## Building Landscapes for Bee Navigation Experiments

I knew that it was necessary to return to the 1920s approach of Ernst Wolf (see chapter 3) and use a locally featureless landscape. But where to find such a setting in the vicinity of Berlin, where I had meanwhile become a PhD student? In 1990, I commuted frequently by train to Hamburg, which involved traversing the German Democratic Republic (GDR) in its final days before German reunification in October of that year. Because of the socialist land reforms in the GDR, multiple smaller farms had often been collectivized into huge "agricultural production cooperatives" (LPG or *Landwirtschaftliche Produktionsgenossenschaften*), and while gazing out of train windows I noticed how this had resulted in horizon-to-horizon flat and featureless fields. Perfect bee-testing territory!

Next, we needed some large artificial landmarks whose positions we could control and that could be rapidly moved as part of experimental manipulations. During an inebriated discussion with my friend and fellow PhD student Karl Geiger, we decided that the ideal shape, in terms of stability, would be a kind of three-footed tent that we could quickly fold up and transport. We didn't have any specific funding for this project, so we had to be creative. We found an East German company that had until recently provided bunting for socialist marches and was glad to find a new customer, even for relatively little money. Our supervisor's wife, Mechthild Menzel, generously stitched the material into tetrahedral pyramid shapes, and we draped each of these over three 4 m long aluminum poles (figure 6.3). We had our large movable landmarks (each about 3.5 m high), and so we were able to test the influences of landmarks on bee navigation under realistically large-scale conditions in a controlled manner.

Finally, we needed the person power to move landmarks, and to count and catch bees. Help came from curious undergraduates at the Free University of Berlin, as well as from my mother and brother. I asked *everyone,* including my (then) eighty-two-year-old grandmother. She said no. I eventually accepted her position, but only after a lengthy discussion. For the summers of 1991–1993, for two weeks at a time, a motley crew of aspiring science students, plus their friends and relatives, camped in barracks on a sports ground five kilometers from the experimental field, sleeping on bunk beds that—so we were told by local officials—had been scrapped by a GDR prison. The price for the accommodation: one German mark (half a euro) per night per person. Before East Germany was colonized by West German business sharks, the prices were fairly reasonable.

We were first interested in the extent to which landmarks, in addition to the sun compass, might be used in direction finding. Thus, we set up four tetrahedral tents in a single file from the hive, at distances of 75, 150, 225, and 300 meters (figure 6.3). Honey bees were trained to visit a feeding station halfway between the third

**Figure 6.3. Artificial landmarks for experiments on bee counting and navigation.** 3.5 m high tents provide movable landmarks for studies of counting and of landmarks' role in direction and distance estimation. Bees landed earlier when more landmarks were encountered en route, and flew farther if the number of landmarks was reduced relative to training.

and fourth landmarks. Then we progressively increased the discrepancy between the learned compass direction and the landmarks, by displacing the entire line of landmarks to increasingly larger angles. We found that the greater the discrepancy between the trained compass bearing and that indicated by the landmarks, the fewer bees "trusted" the displaced landmarks, and at 30 degrees of discrepancy, they ignored the landmarks altogether when they had access to their sun compass. Under heavily overcast conditions, when neither sun nor a polarization pattern of the sky was available, more bees followed the landmarks—but again, less so when the landmarks indicated a direction markedly different from the learned direction. In this case it is possible that they used a magnetic compass.

Clearly, in both types of weather conditions, bees had followed their learned vector instruction ("fly 187.5 meters south"), and used landmarks only for fine-tuning of the course. The same relationship was found when we introduced discrepancies between a unique landmark (made up of three differently colored tents) at the feeder and the learned distance. Bees largely ignored this landmark when it was located far "off course," but gave credence to it when it was closer to the remembered location. In short, bees "on a mission," flying to a known destination, are guided principally by a flight vector, and will use landmarks as a backup system for fine-tuning the course. This is useful for correcting errors during navigation, such as those that might arise due to passive displacement by wind.

Our arrangement of a series of four identical landmarks also lent itself to the question of whether bees might be able to count the landmarks. During training, bees had flown past three landmarks between the hive and a feeder. What would happen if we produced a contradiction between the correct memorized distance and the number of landmarks—if, for instance, the bees encountered four or five landmarks within the identical distance to their intended target? While the memory for the trained flight vector was strong, we found that the more landmarks the bees encountered in the familiar direction, the earlier they tended to land. When, conversely, we reduced the number of landmarks between the hive and the habitual feeder from three to two, many bees overshot the trained distance to land after the third landmark (now beyond the training feeding station). At the time of publication in the mid-1990s, many scholars were skeptical about the possibility of counting abilities in insects, but these abilities have since been replicated in other experimental paradigms, in honey bees as well as other species of bees.

Interestingly, counting may rely on different behavioral strategies in insects than in humans. We can identify small numbers at a glance, in a strategy called "subitizing." For example, you can

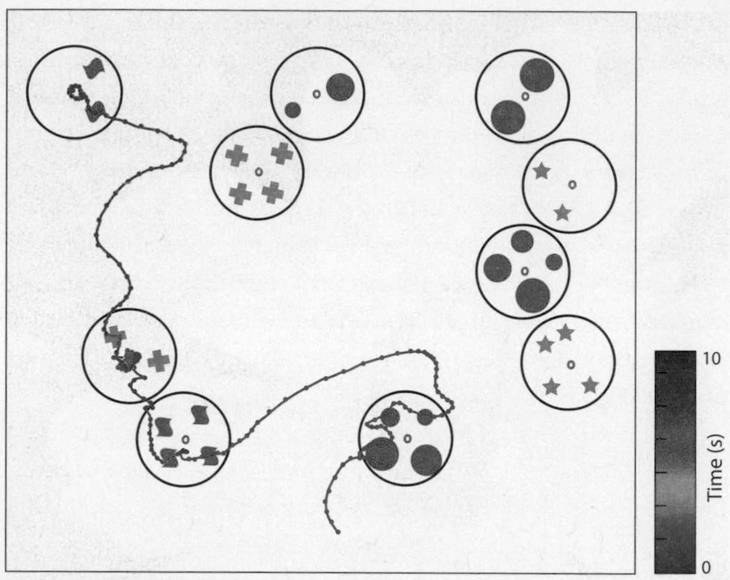


**Figure 6.4. Sequential tagging of countable items by a bumble bee.** Flight path of a bee trained to select stimuli with two items and avoid those with four. The first 10 s of the bee's scanning behavior are shown; the path is color-coded to show the progression from early (violet) to late (red). The bee sequentially examines two patterns containing four items, but rejects each of them after scanning three items in each. She then chooses a pattern containing the correct number of two purple crosses (even though she has not been rewarded on any other items than yellow ones before) and finally selects another pattern with the correct number of two (yellow) dots. Dots are separated by time intervals of 33 milliseconds.

identify the number of dots on a die extremely quickly. Bees, conversely, appear to have to "tick off" each item sequentially, in a strategy analogous to humans pointing a finger at each countable item in turn (figure 6.4). It remains to be determined whether, with extensive training, bees could also learn to count small quantities by parallel processing, and thus more rapidly.

Recent studies hinted that the numerical abilities of bees might be highly advanced, purporting to show that bees can add and subtract, and even understand the concept of zero. However, as with many other studies on animal counting, it is at present not fully clear whether bees used number or some alternative cue to solve these tasks. This is because stimulus number is often correlated with other, non-numerical cues, such as the area covered by the countable items, contour length (the total boundary length of countable items), or their "convex hull" (the shape connecting all the outer elements). All such alternative explanations must be carefully excluded by experimental design.

## Path Integration in Bees

*Path integration*—also called "dead reckoning"—is a behavioral ability to move straight home from any point in an animal's home range, even if the home is not visible from that point, and even if the path up to that point was a searching trajectory with many loops and turns. This requires that the animal continually update its position relative to home (like holding a mental elastic band that ties it to the nest) by integrating the angles turned and distances traveled (figure 6.5). Path integration had long been studied in desert ants, especially by Rüdiger Wehner (see chapter 3) and Tom Collett (see the section above, "Learning in Context") and their teams. Deserts are fruitful natural laboratories for such tasks, since they offer terrains entirely free of visual landmarks.

Up until the 1990s, the only evidence for path integration in bees was indirect. It came from a study by Karl von Frisch, in which he had trained bees to fly back and forth between their hive and the feeder along a triangular route around the edge of a rock

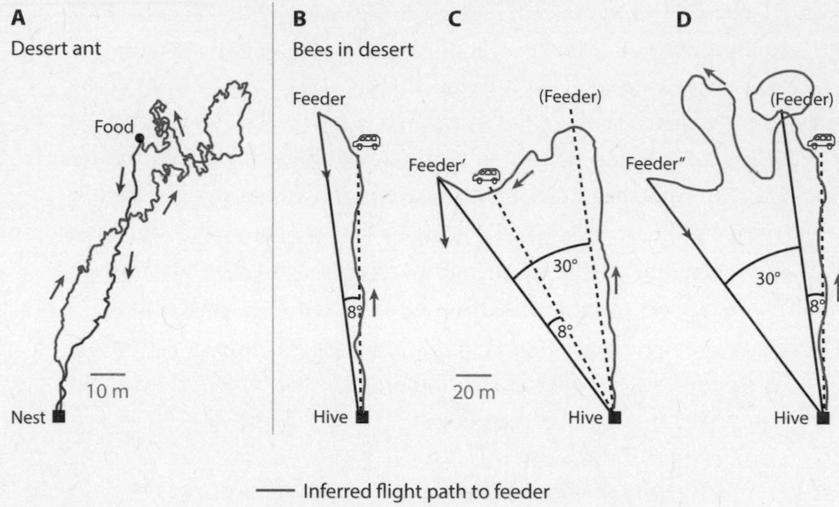


**Figure 6.5. Path integration in desert ants and honey bees.** ***A.*** An individual desert ant searches for food along a convoluted track of over 350 m (green). Once food has been found she does not retrace her steps but instead heads straight home (red). ***B.*** Honey bees are trained to fly to a destination 175 m from their nest via a landmark (car symbol). ***C.*** When landmark and feeder are displaced by 30° (food now at Feeder'), bees do not use path integration, but instead take the familiar direction from the feeder. ***D.*** When the landmark is in the familiar position, but bees have to locate the feeder in a new location (at Feeder"), path integration takes them onto a flight straight home.

spur (bees did not solve the problem by flying over the top of the mountain). Despite the circuitous flight, bees returning from the feeder indicated in their dances a route directly to the feeder on the other side of the cliff—as if indeed they had flown straight through the mountain. Apparently, like dead-reckoning desert ants, the bees calculated a direct route to the feeder by integrating the geometry of the individual flight segments, but direct evidence of path integration was still missing.

In 1994, I visited my friend Jan Kunze (1968–2021) in Tucson, Arizona, where he was at the time doing experiments for his master's thesis. Two driving hours east of Tucson, we found a completely featureless, flat sand desert in Cochise County, called Willcox Playa. Jan had already worked with me on the "tent experiments" in northeast Germany described above; we took inspiration from Wehner's and Collett's work on desert ants, and decided to do a path integration experiment with honey bees in this desert. We borrowed a honey bee hive and a car from friends at the US Department of Agriculture in Tucson. Remarkably, even though no honey bee had ever likely flown in this wholly featureless and vegetation-free terrain, the bees coped well and managed to navigate successfully between their hive and a feeding station we set up.

In those days, there was no technology to track the bees' flight; path integration had only been studied in walking animals, whose paths through complete foraging bouts are more easily observed. In bees, all we could do was record when and where they landed, and their "vanishing bearings"—a technique borrowed from studies of pigeon homing, where a flying animal is observed for as a long as possible from a departure point, and then its compass bearing is noted at the moment it disappears from sight.

We first trained bees to a feeding station 175 m north of the hive, close to a landmark (which, for want of anything else, was the USDA vehicle). After training, we removed the feeder from the familiar location and presented a new one 30 degrees to its left (at the same distance from the hive). Bees initially searched near the familiar location, but after further searching, several individuals discovered the new feeder location. On that initial visit, when the foragers had fed to satiation and embarked on their way home, we recorded their vanishing bearings. Even though the bees had not flown directly to this new feeder, and even though the hive was too far away for the bees to see, they nonetheless flew straight in the direction of the hive from this new location. Thus bees, like ants, use path integration for homing from a novel destination, even when no local landmarks indicate the direction of home. Presumably, they integrate information about the distance they've flown (measured by *optic flow*, or how the landscape rolls by underneath during forward flight) and the angles they've turned (measured by the sun compass).

Interestingly, bees did not slavishly follow their path integrator, but used it only when it made sense. When the landmark was displaced together with the feeder by an angle of 30 degrees to the left of the training station, bees behaved as if there had been no displacement, and flew in the familiar direction from the previous feeding location (figure 6.5). It thus appeared that if the feeder was in the expected position relative to the landmark, bees attributed the fact that they had arrived there by a circuitous route to their own navigation error (or perhaps displacement by wind), disregarded their path integrator, and simply retrieved the familiar home vector from the feeder at its new location.

It might well be that flying insects such as bees need an extra level of flexibility in the use of path integration compared with walking animals such as ants. After all, when you are in constant contact with the ground, how far you have gone and in what direction is entirely in your control. The same is not the case for flying animals, which are subject to passive displacement by wind. For them, familiar landmarks and locations offer a more reliable indication of the correct direction home than does any evaluation of one's own recent movements. Interestingly, path integration in bees appears to require visual input: bees walking to feeding stations in complete darkness can measure distances and directions accurately (most likely using idiothetic, or internal, cues based on proprioception), but fail to find the correct direction home from a newly discovered feeder by path integration.

In recent years, the neural mechanisms underlying path integration have been explored in depth. Compass neurons (analyzing flight direction based on the sun's position and / or on polarized light) and speed-encoding neurons (measuring flight distance based on optic flow) converge in the so-called central complex of the insect brain (see chapter 9). This structure contains all the neural circuitry required for steering via path integration. Based on this information, elegant and comprehensive neural models have been developed that capture the entire pathway from visual sensory input to behavioral output. Predictions from these models have been tested on a (wheeled) robot that can successfully display insect-like path integration. Such models and robotics have yet to incorporate the cognitive flexibility found in flying bees, which can selectively turn off their path integrator when landmark information indicates that their vector integration has yielded erroneous results.

## Radar Tracking of Bees

Our long-distance bee orientation experiments in the 1990s (as well as preceding ones from the decades before) suffered from the same shortcoming. We never knew what bees did between the moments when they vanished from one known observation site and reappeared at another. When displaced, at what point did they realize that they were in an unexpected location and start searching? What search strategies did they use? What familiar landscape features did they recognize when searching, and did their recognition result in a straight flight home? Could bees "change their mind" somewhere midway along a route and alter their course? We simply could not answer any of these questions, since no transmitters

**Figure 6.6. Harmonic radar technology for insect tracking.** ***Left:*** The harmonic radar transmitter (bottom dish) sends a microwave signal to the transponder (see right); the transponder translates the signal to the 2nd harmonic (twice the original frequency) to send it back to the receiver (top dish). ***Right:*** Bumble bee worker wearing transponder.

were light enough for the bees to carry so that one might track their long-distance flights.

But just one year after our above-described bee orientation studies were published (in 1995), a revolutionary technology was developed by a team of engineers and biologists at Rothamsted Research: harmonic radar. This technology does not require bees to carry a battery-driven transmitter, only a 15 mg device called a transponder, which is attached to the back of a bee and is much lighter than the nectar load a bee can carry (figure 6.6).

This technology has made it possible to follow bees' spatial whereabouts for their entire lifetimes, from the first moment they ever leave the native nest through their early wanderings while exploring their surroundings, and their later navigation to discovered flower patches, until their death a few weeks later. We were interested in finding out how, in a natural setting, bees would divide up their foraging career between landscape and resource *exploration* and *exploitation*. The first flights of bumble bees and honey bees are invariably orientation flights, beginning near the nest or hive, where bees initially behave as described by the pioneering Polish apiarist Johann Dzierzon (1811–1906) at the dawn of the twentieth century (see the opening quote of this chapter). First, flying in loops of increasing size near the colony, often while facing its entrance, bees memorize the appearance of the nest and nearby landmarks. Eventually, they leave the vicinity of the hive, and explore the further surroundings up to several hundred meters away from the nest in large loops, often in various directions from home. In honey bees, the sole function of such flights is to acquire spatial information; they do not appear to visit flowers yet on their first flights. In contrast, bumble bees on their first flights—which can last over two hours—not only do spatial reconnaissance but will often already probe their first flowers.

In one case, we were able to follow a bumble bee worker for 156 foraging bouts, until it vanished from the radar during a regular trip on the thirteenth day of its outdoor foraging career (figure 6.7), presumably taken by an insectivorous bird or a crab spider. It's instructive to examine this bee's "life story" in some detail, because it allows unprecedented insights into how an animal changes its spatial movements throughout its life.

On its first day, the bee made only two excursions. The first flight lasted two hours and 18 minutes—the longest this bee would ever take. The flight described a number of loops, between which she periodically returned to the vicinity of the nest, without entering it. The loops took the bee in multiple different directions, including to a forest edge north-northwest of the nest (which will become of interest later). This orientation flight covered a large area in almost all directions from home, except southwest. On the following morning, the bee made another 77-minute

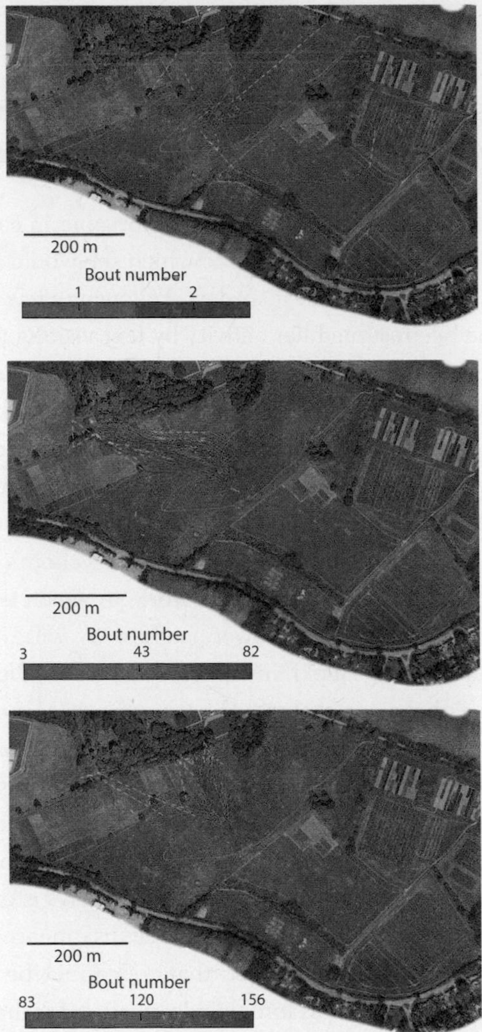


**Figure 6.7. An individual bumble bee tracked from its maiden flight to its very last foraging bout.** In all panels, the color green denotes early flight activity, and various shades via yellow, orange, and red indicate later activity. ***Top:*** The first orientation flight of the bee lasted over 2 hours and included multiple sequential loops in various directions, and several returns to the vicinity of the hive. ***Middle:*** After another orientation loop at the beginning of the 2nd day, the bee then found a suitable foraging patch which she continued to visit for several days and dozens of foraging bouts. ***Bottom:*** After an interruption, the bee briefly returned to the previously exploited patch before spending the rest of her days foraging from a second foraging patch.

orientation flight, consisting of loops mostly toward destinations west and southwest of the nest, an area she had left less explored on the first day.

On her fourth flight, the bee began foraging in earnest—she had discovered a patch of flowers to which she would return several dozen times over the next six days. Following a few days of bad weather, the bee resumed her activity by first visiting the familiar patch on day 9, but then, on day 10, "changed her mind" halfway along a flight to this patch and flew instead to a different location she had explored only once, nine days earlier, during an orientation flight. She then visited this location exclusively for the rest of her life until her sudden disappearance on day 13.

This is a relatively simple bee life story: a few hours of youthful search followed by a lifetime of hard work, in which the bee only ever appeared to exploit two flower patches. We will later see that not every bumble bee's life is this simple (chapter 10), but nonetheless, this bee's behavior captures the key elements of exploration and resource exploitation in an instructive manner. It is also interesting that this bee, upon abandoning her first foraging patch, did not perform further reconnaissance flights, but appeared instead to fly straight to a different patch (north-northwest of the nest) that she had only visited once, nine days previously, on her first flight ever. Unfortunately, the bee disappeared from the radar for part of that journey, so that we cannot be certain by which path she reached this foraging destination for the first time. Nonetheless, the suggestion that a bee can "change its mind" while en route to one destination and choose another memorized one instead is interesting, because it relates to the idea of the cognitive map, whereby an animal can "imagine" multiple familiar destinations and compute new shortcuts between them. However, observational data such as these reported above cannot provide a decisive answer to this question.

## Exploring the Cognitive Map with Jet-Lagged Honey Bees

The availability of harmonic radar technology also prompted my former supervisor, Randolf Menzel, to reexamine the question of whether bees might have a cognitive map. Departing from his stance in 1990, he published a series of papers in the 2000s that contained claims that bees did, after all, possess such a map. But what is the evidence for cognitive mapping in bees?

Menzel took his team to the same large flat field in the Brandenburg region of Germany where we had done our work a dozen years earlier, and set up a little landscape of the tetrahedral tents we had designed back then. It had been known since Fabre's work in the nineteenth century that bees can return home from distant locations from which their home cannot be seen. But harmonic radar made the entire trajectory of bees after such displacements visible for the first time. Radar tracking confirmed the three flight phases that had been postulated by Wolf in the 1920s: honey bees displaced from feeding stations and released at a different site would initially behave as if they had not been displaced at all—they would fly the same distance and compass direction as if nothing had happened. Near the end of their memorized flight vector, things got interesting: bees slowed down their flight and began circling in loops, as if searching for familiar landmarks that might indicate the way home. And typically, after a few such loops, bees would apparently recognize the sought-for landscape features, and then fly straight to their hives. But are such straight homebound flights proof of a cognitive map?

A decisive indicator of such a map is that it allows the animal to perform novel shortcuts between familiar locations—by combining memories from separate routes. To prove that the animal really has "thought up" a new route, it must be shown that the route really is novel. The animal must neither have familiarity with the route from previous experience, nor must it be able to just see the target. And since the honey bees had not been tracked continuously for their lifetime, it was impossible to ascertain if the tested animals had not traveled similar routes before. Indeed, as we have seen, even a single orientation flight can take a bee in all cardinal directions from the hive. In addition, we had already learned in our 1990s studies that bees can tie flight vectors to familiar landmarks. It is therefore possible that bees, from the moment they recognized any familiar landscape features during their search, retrieved the appropriate homebound vector (at which point they display straight flights to the hive without the need for a mental map).

Menzel and co-workers came up with a nifty trick to disable the most obvious component of such vectors—the flight direction, as computed using the sun compass. Since use of the sun compass requires bees to know the time of day, any disruption of their sense of time should result in miscalculations of their homebound vectors. Using the general anesthetic isoflurane, normally used to knock out human patients for operations, the authors put their experimental bees to sleep for six hours, suspending their circadian clocks for that duration. This essentially produced jet-lagged honey bees: it caused them to "think" that the sun was still in the east (morning) when it was really already in the west (afternoon).

As predicted, these bees made huge but consistent directional errors during their initial departure bearings when displaced, using their sun compass. However, when these time-shifted bees found themselves at an unexpected location near the end of their trained flight vector, after again looping around in search of familiar landmarks they still embarked on straight flights home (which, in this case, they could not have done by attaching a flight vector—memorized relative to the sun—to a familiar scene). But, as other authors have pointed out, the bees might also have simply moved to reduce the mismatch between the currently viewed scene and a memorized one, and doing so (perhaps several times along the way) would have taken them ever closer to the hive.

The question of whether bees navigate by cognitive maps thus remains open. It would be useful to return to the key criterion of whether bees can determine novel shortcuts, along routes they've demonstrably never traveled before. It would also be worthwhile to move away from homing—the home is such an important center in a bee's life that it typically dedicates extensive efforts during its orientation flights to ensure that it can return home reliably from all directions. The really interesting spatial challenges arise when bees have to juggle memories of multiple spatial locations, as they do in the case of flower patches.

## How Smells Bring Back Distant Memories

In volume I of his novel *À la recherche du temps perdu,* Marcel Proust describes how the narrator, after tasting a tea-soaked madeleine, suddenly recalls long-lost childhood memories in vivid detail, of a time when his aunt had fed him the same kind of biscuit. In an experiment that resembles this narrative, Mandyam Srinivasan, an Indian-born Australian biologist, and his team conducted ingenious experiments in which they explored whether bee memories could be similarly triggered by the smell of food.

To explore the accessibility of memories of out-of-sight destinations, the team first trained honey bees to two feeding stations, one of which was rose-scented and the other lemon-scented, giving the bees exposure to both feeding stations. Subsequently the experimenters blew one or the other scent into the hive—and sure enough, the bees arrived at the correct feeding station, which wasn't scented during the test. The bees' spatial memories had been triggered by smelling the familiar scent, much as Proust's protagonist recovered distant memories when exposed to a familiar taste. These observations of Srinivasan and co-workers are also reminiscent of Lindauer's tentative observation that bees can spontaneously recall spatial memories while inside the hive—even in the depth of night.

## Bees and the Traveling Salesman Problem

The number of flower locations that must be remembered by bees can be much higher than two, since they typically have to visit many flower locations (often distributed over a wide area) to fill their honey stomach even once. In doing so, bees face a challenge equivalent to the so-called traveling salesman problem: the need to link multiple destinations in a sequence that minimizes travel distance and time. When I came to Stony Brook as a postdoctoral researcher in the 1990s, my mentor James Thomson was interested in how bumble bees cope with such challenges, and had done elegant studies of how bees remembered routes between multiple flowers, both in laboratory settings and within natural flower patches. Exposure to his work prompted me to return to my early interests in sequence learning in bees, and harmonic radar technology allowed us to explore how bees cope with the challenge of multi-destination routes over realistic field scales, when the next flower or patch is not easily visible from the current one.

In a flat field north of London, we set up five feeding stations, arranged in a regular pentagon (50 m to a side). The traveling salesman problem becomes rapidly more complicated with an increasing number of locations to link: for three sites, there are just six possible routes ($3 \times 2 \times 1$), but for five sites there are 120 ($5 \times 4 \times 3 \times 2 \times 1$) possible ways to link the stations. Indeed, for bees the problem is even more complicated than for human salesmen, since bees do not have a map at hand to start with: they must first discover the locations one by one by individual exploration. Would bumble bees nonetheless find the best of 120 options?

We discovered that bees typically established stable routes linking all the feeders in an optimal sequence after just 26 foraging bouts, during which bees tried only about 20 of the 120 possible routes. Radar tracking of selected flights revealed a dramatic 80 percent decrease in total travel distance, shaving off about 1,500 meters between the first and the last foraging bout. When a feeder was removed, bees continued to inspect the obsolete location for a considerable time. Bees also engaged in more-localized search flights when a previously known feeder was missing. This strategy facilitated the discovery of new feeding stations and their integration into a novel optimal route.

The bees' gradual optimization procedure appeared to be based on trial and error—nothing indicated that a cognitive map provided bees with a sudden insight to compute the optimal route. Instead, bees had a consistent tendency to experiment—to try new ways to link the feeding stations, and if a novel route provided an improvement, to switch to that one. Even highly experienced bees that had already found the optimal route continued to experiment with novel solutions intermittently—a strategy that is likely to be successful not only in optimizing routes in stable arrays, but also for discovering new feeding locations when flower profitabilities and locations change.

We have learned that bees are tremendously adept at navigating between multiple locations—but it was the murderous parasitoid (and solitary) ancestor of today's social bees that "invented" the building of a nest for its offspring, necessitating the evolution of a highly precise spatial memory and rapid learning. The brain structures that evolved to enable such learning would in turn have facilitated precise spatial learning about food sources. It is likely that these learning abilities also came in handy for memorizing the features of floral food sources, which we will explore in the next chapter.

# 7

# Learning about Flowers

That insects should visit the flowers of the same species as long as they can, is of great importance to the plant, as it favours the cross-fertilisation of distinct individuals of the same species; but no one will suppose that insects act in this manner for the good of the plant. The cause probably lies in insects being thus enabled to work quicker; they have just learnt how to stand in the best position on the flower, and how far and in what direction to insert their proboscides. . . . They act on the same principle as does an artificer who has to make half-a-dozen engines, and who saves time by making consecutively each wheel and part for all of them.

**—Charles Darwin, 1876**

A pollinator flying through natural habitats typically encounters several dozen plant species in flower, all differing in their advertising billboards and in the quality and quantity of the rewards they contain. In the previous chapter we explored bees' extraordinary memory for multiple flower locations, and earlier we learned, from von Frisch's work, that bees can learn to associate the colors of flowers with the rewards they contain, using these and other sensory cues, such as scents and electrostatic fields, to predict the best foraging prospects. In this chapter we will discover how bees integrate information from multiple cues, direct their attention to single out the relevant ones, learn rules to classify flower types, and juggle multiple memories of how to handle flowers.

## Learning to Handle Electronic Flowers

While some flowers allow relatively easy access to pollen and nectar, these are often not the ones with the most impressive rewards. But some plenteous flowers, such as snapdragons (*Antirrhinum* spp.) and monkshoods (*Aconitum* spp.), require bees to perform some rather advanced contortions to get at their bounty (figure 1.4). In fact, such flowers are natural "Skinner boxes"—the classic experimental puzzle boxes in which rats or pigeons learned by trial and error that certain actions would lead to rewards (or punishments).

My postdoctoral mentor James Thomson and I wished to explore how long it takes bees to figure out the right movement sequences to gain access to the sweets in flowers, and whether bees' memory had the capacity to retain multiple such movement sequences—different ones for each flower species. To test learning and memory rigorously, one must work with animals whose experiences prior to an experiment are known; this is impossible with freely flying honey bees and other animals simply observed in the field. But James Thomson worked with bumble bees, and knew how to raise them from queens caught in the wild. Together, we thought that it might be feasible to test them in flight arenas within a laboratory, where we could control the experiences of the bumble bees before and during experiments (figures 7.1 and 7.2).

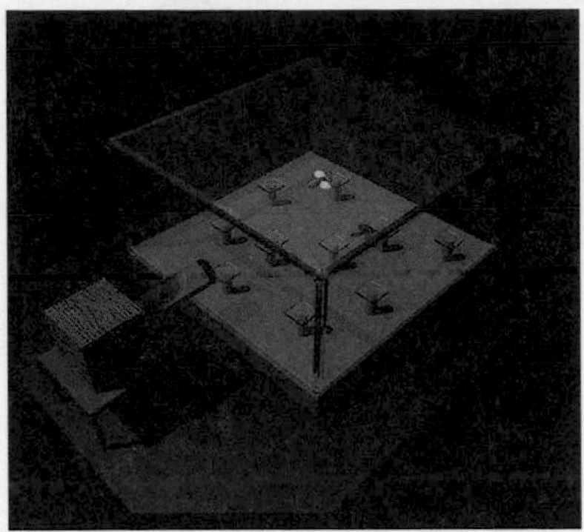

**Figure 7.1. Experimental flight arena and bumble bee with artificial flowers.** ***Left:*** Bumble bees can be kept in nest boxes connected to flight arenas in which artificial flowers of various types are offered to the bees to forage from and to discriminate among based on profitability. ***Right:*** A number-tagged bumble bee worker inspects an artificial flower. This design allows combining a visual pattern with a scent delivered through a plexiglass tube.

We were pleased to see that worker bees would confine their entire foraging to a very manageable flight arena. So long as bumble bees are offered the commodities they crave, nectar and pollen, they will shuttle back and forth between their colony and the flight arena. For as long as the experimenter is willing to keep up with the stamina of the bees, bees will keep returning to the arena to refill their honey stomach. (Once I observed a single bee foraging for 16 hours, at which point I, not the bee, gave up). Once full, worker bees come back to the nest and regurgitate their food, only to return a few minutes later to collect more goodies. We knew we were ready to start some experiments in earnest.

To do such experiments in a controlled manner, one cannot work with wild flowers. They vary in reward levels and morphology, and the scent marks that visiting bees leave on them

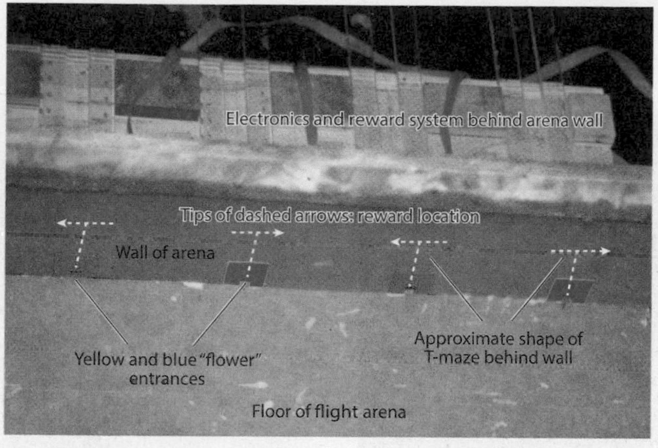


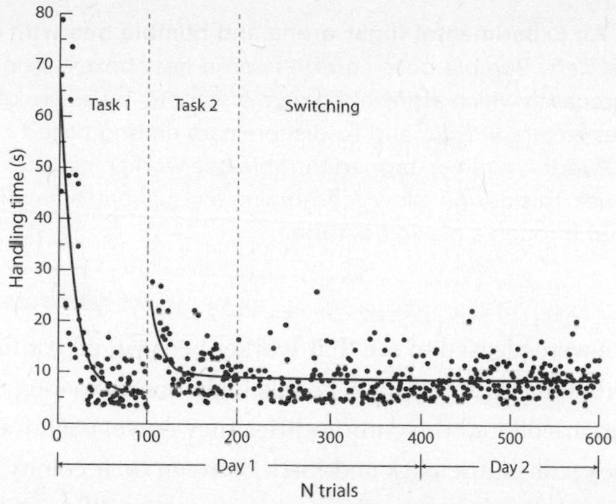


**Figure 7.2. T-maze "flowers" and a single bumble bee's proficiency at handling two types of flowers.** ***Top***: Experimental flight arena to study bees' learning of flower handling procedures. Bees had to learn that if the entrance of the flower was yellow, they had to turn left inside the flower, and they had to turn right if the entrance was blue. ***Bottom***: Flower handling times in a T-maze flower for a single bee as a function of the number of trials (flower visits). The bee first learned to associate blue with right turns (first 100 visits), then yellow with left turns (second 100 visits), and then for 200 visits had to switch between both tasks. A further 200 such visits took place on the second day. Note the transfer effects from task 1 to task 2: the bee performed better initially on task 2 than on task 1, but its saturation level on task 2 was worse than on the task learned first (indicating interference). The bee stayed at this performance level when forced to switch between the two flower types; no overnight forgetting was apparent.

are impossible to remove. So James Thomson built some magnificent plastic flowers—tiny T-mazes equipped with a mechanism that would refill a precise volume of sucrose solution after every bee visit (figure 7.2). The flowers had a color-coded entrance: blue meant that the bee had to turn right inside the flower to find the reward, yellow meant it had to turn left (or vice versa). In this way, the two movement sequences were of the same difficulty, but reversed.

For automatic recording of the bees' accuracy and speed, I built three infrared light barriers into the flowers, one at the entrance and one in each arm. The electronics were mostly soldered together from components plundered from the fischertechnik® construction toys I played with as a child; I used my old Lego® bricks, too, as nectar feeders inside the bee colonies. I wrote the Turbo Pascal software that interfaced the light barriers with my computer so that each time a bumble bee broke a light barrier, a tone sounded—a different one for each light barrier. This allowed me to monitor whether the electronics were working accurately, but also, since each of the many light barriers generated a different tone, the behaving bees made some very interesting music.

Bumble bees learning to handle the artificial flowers were initially as clumsy as naïve bees are with natural flowers: it took them five to ten times as long to extract the reward as it did an experienced forager—sometimes they needed a full minute to exploit a flower. However, they typically improved rapidly over the next few dozen visits, and then reached a constant level of efficiency. Learning speed was not dependent on whether a reward was present every time, but on the number of times bees performed the correct motions. When bees learned to handle a novel flower type whose morphology was similar to one previously learned, *positive transfer* occurred: bees were able to carry over some of their existing expertise and did not start out on the new species as poorly as an entirely naïve forager did (figure 7.2, bottom). Bumble bees learned to forage from artificial flowers with two distinct morphologies, although with slightly elevated handling times and handling error rates, compared to bees just learning to handle one flower type. There is no question that, like human assembly line workers, bees can technically perform more than one task—though efficiency is somewhat lower if multiple tasks are performed in alternation. Once bees had learned flower handling procedures, they were retained in memory for more than three weeks (the lifetime of a worker bee), although some decay occurred. It seems that bees never truly forget motor memories. The same is true for humans, as you'll find out if you try swimming, ice skating, or skiing after an interruption of many years.

## How Bees Pay Attention to Flowers

In most animals' environments, the amount of information perceived by the senses exceeds the brain's information-processing capacity by several orders of magnitude. In 1997, Johannes Spaethe became my first PhD student, and he explored the question of how bees selectively focus only on certain familiar flowers, while their sensory system is bombarded with stimuli from all the other flowers bees see as they fly through meadows. Attention is a kind of "inner eye" that allows animals to focus selectively on different aspects of information coming from the sensory periphery. Consider the "cocktail party effect," which allows you to focus on a single voice out of many talking in a room; or the young mother who wakes at the slightest sound of her newborn while ignoring all other noises, even much louder ones. But do bees pay attention? This is not a trivial question, since paying attention implies that one knows what one is looking for—that something is in your mind before you actually see it. To understand the significance of attention in the search for flowers, we need a small excursion into the bees' spatial vision.

Bee eyes are composed of several thousand functional units, the *ommatidia*, each containing its own lens and set of photoreceptors. Each ommatidium corresponds to one "pixel" in the bee's visual system—so the optics of the bee provide only a relatively coarse-grained image (figure 1.2). Since the bee eye is curved, each ommatidium looks in a different direction (about 1 degree difference between adjacent ommatidia). But the spatial resolution of bee vision is limited not only by the interommatidial angle, but also by subsequent neural processing. The receptive fields of color-coding nerve cells in the bee brain are very large, meaning that they pool the signals from several dozen neighboring ommatidia. The result is that bees don't see color very well from a distance: from one meter away, a flower must be enormous (26 cm in diameter) to enable a bee to either recognize its color or detect it using color contrast.

But bees are able to use a different neuronal channel with a smaller receptive field when they are farther away from a flower. When a flower is seen with an angle subtending at least 5 degrees (meaning that if you draw a triangle between the flower's edges and the bee eye, the angle on the bee eye is 5 degrees, and becomes larger as the bee comes closer to the flower), honey bees employ a monochrome signal for detection: that is, the difference in signal provided by the green receptor between background and target. This trigonometry still means, however, that a honey bee must be no more than 11.5 cm from a 1 cm diameter flower to detect it. This severely constrains the rate at which flowers can be found. Accordingly, search time increases steeply with decreasing flower size.

## A Gear Box in the Brain of Bees

This two-channel flower detection system might mean that, as the bee searches for flowers in a meadow, the monochrome channel would always kick in first (since it can see flowers from a greater distance, and it's also faster: it responds in under 8 milliseconds), and the actual color would only be discernible when the bee has approached much closer (the color channel, as well as more coarse-grained, is also slower; the UV receptor, the slowest of the three, takes more than 12 milliseconds from stimulus onset to build up an appreciable response). But obviously, the green receptor signal is much less accurate in identifying a flower than is the complete trichromatic system. So do bees really use these two channels passively, always the low-accuracy, high-resolution channel first in the approach to a flower, and then the high-accuracy, low-resolution channel second? If so, bees might waste a lot of time approaching arbitrary flowers or other colored objects, only to discover when they came close that they had the wrong target.

Johannes Spaethe wished to test whether bumble bees really used such a simple but possibly maladaptive strategy in the search for flowers. He set up plastic flowers in a flight arena and precisely monitored the bees' flight behavior as they searched for the flowers, which were shuffled to a different location in the arena each time the bee returned to the nest with a full tummy. He started with relatively large plastic flowers (28 mm in diameter) and then progressively made them smaller and smaller (down to 5 mm in diameter). Initially the bees ignored the information from the monochrome, high-resolution, high-speed channel, and the flowers' detectability was directly determined by the color contrast with the green arena floor. As flowers became smaller, bees flew more slowly and also closer to the ground to facilitate detection. However, to detect the smallest flowers using their sluggish color detection system, flight would have to be so slow as to be no longer profitable (in fact, 25 times slower than when surveying the largest flowers). In this case, bees switched to their low-accuracy, but high-speed monochrome channel. This meant that bumble bees, rather than responding passively to whichever signal entered their visual system first as they approached a flower, suppressed the monochrome input—except when searching for the smallest, hardest-to-detect flowers. Bees essentially have an attentional gear box: depending on environmental need, they know to use the high-speed, high-resolution, but low-accuracy gear (the monochrome input) or the high-accuracy, but slow and poor-resolution gear (which processes the color signal). More generally, Johannes's study showed that the movement speed of the bees when searching for flowers was not constrained by the biophysics of flight, but by the brain's processing speed.

## How Much Information Can a Bee Process at a Glance?

Johannes then asked how attention might constrain the search for flowers where there are multiple types of flowers in a scene. Do pollinators seeking out flowers of a particular species (while ignoring others) process all the stimuli that they encounter by means of so-called parallel processing, or by serial processing? If information processing is serial (i.e., one "bit" of incoming information is analyzed at a time), then the efficiency of finding a target flower will be constrained by how many other items ("distractors") are simultaneously present in a scene. If, however, processing is parallel, flowers of multiple species can be examined simultaneously.

Johannes discovered that in honey bees, the search for visual targets that differ in color was strictly serial. This means that the accuracy and speed with which honey bees found a target depended on the number of distractors that were simultaneously presented in the target's vicinity. This contrasts with humans, who can examine stimuli in parallel if targets and distractors differ only in one stimulus dimension (such as color or shape). The target is said to "pop out," and search time and accuracy are unaffected by the distractors that are also present in the scene. If honey bees are indeed limited to serial searching, this has fundamental implications for flower search under natural conditions. It means that the efficiency with which bees find flowers is constrained not only by qualities of the target flowers (e.g., size, color, and contrast to the background), but also by those of other flowers in the area competing for attention.

Even though bumble bees solved this task more efficiently than honey bees, Vivek Nityananda, a postdoctoral fellow in my team, later discovered that bumble bees are still constrained in their attentional resources in ways that you and I are not. A human can extract critical information from a scene (for example, is an animal present?) at a glance, even if it is flashed on a computer screen only for an instant. However, bees fail all but the simplest discrimination tasks if stimuli are only briefly presented on a screen. They make characteristic scanning movements while learning and recognizing visual patterns, and fail when the opportunity to scan patterns is taken away. The small brains of insects may constrain attentional resources to the point that they can only search for targets by scanning the scene sequentially, rather than by taking in the entire visual field simultaneously. There might thus be very tight links between action and perception in bees, such that when there is no action (no movement of the eyes), there is also little shape perception (figure 7.3).

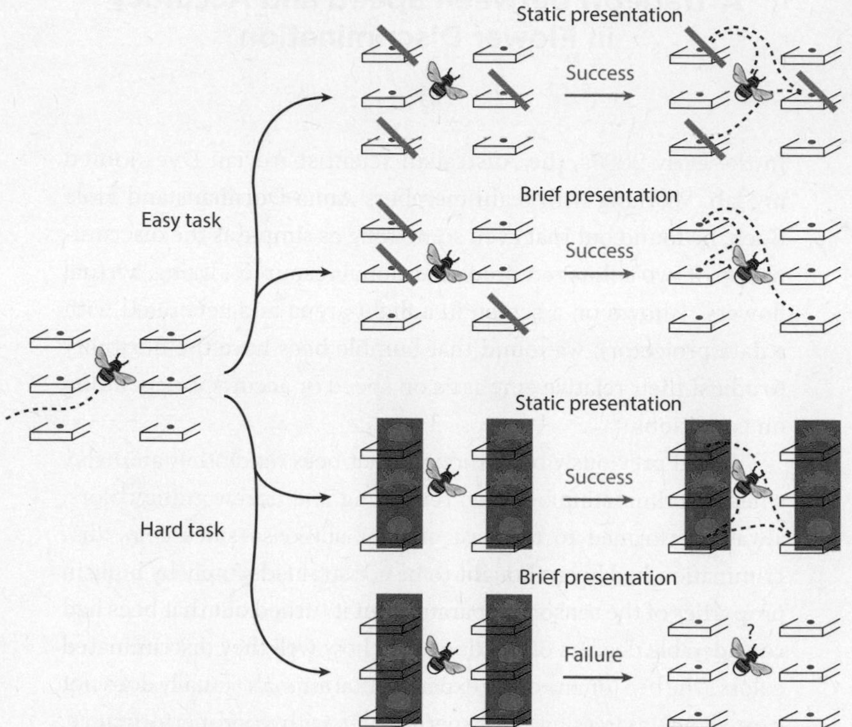


**Figure 7.3. Can bees see at a glance?** A fundamental difference between bee vision and primate vision is revealed when stimuli are flashed at increasingly short durations. While primates can spot salient details in a visual scene rapidly at a glance, bumble bees require a longer view to solve a more complex visual pattern task than they do for a simple task. Bees were trained to six perches in front of a computer screen, three of which offered sucrose solution reward and three of which contained bitter quinine solution, which bees detest. Bees could successfully solve a simple task (locating a yellow diagonal bar marking the sucrose perches) when given either long or brief (25 milliseconds) presentations of the diagonal bars (***upper right***). For a harder task (discriminating a circle marking a sugar feeder from an image of a spider-infested flower), bees could only solve the task when given a static presentation of the visual stimuli (***lower right***), indicating that active scanning was required for the resolution of shape.

## A Tradeoff between Speed and Accuracy in Flower Discrimination

In the early 2000s, the Australian scientist Adrian Dyer joined my lab. Working with team members Anna Dornhaus and Fiola Bock, he found out that even something as simple as the discrimination of two colors required attentional resources. Using "virtual flowers" (shown on a screen in a flight arena and generated with a data projector), we found that bumble bees have the flexibility to adjust their relative emphasis on speed or accuracy, depending on conditions.

It had previously been thought that bees (and other animals), when discriminating between rewarding and unrewarding colors, always performed to the best of their abilities—since color discrimination had been thought to be constrained simply by built-in properties of the sensory apparatus. But it turned out that bees had considerable degrees of freedom as to how well they discriminated colors. The bee (or any other experimental animal) actually does not care about impressing the experimenter with good performance. It cares about obtaining the rewards as fast as possible. When the penalties for making errors are nonexistent or small, then making errors actually can be the cleverest strategy.

Imagine a multiple choice test in which you have to tick off correct statements as correct—but there is no penalty for ticking off incorrect statements as correct. In that case, the fastest strategy to get maximal points is to not read the statements at all and just tick everything off as correct. In the same vein, if you are giving an animal one rewarding color and one unrewarding one, but the colors are so similar that it takes substantial inspection time to make a choice, then it's quicker to be quite sloppy in your color discrimination and simply choose all the options—you'll get your rewards anyway. And it turned out that's exactly what bumble bees do. If, however, penalties are introduced for choosing the wrong option—Adrian chose bitter quinine solution to fill the alternative flower (versus sugar solution in the target flower)—bees' performance suddenly improved tremendously. Decades of work in which claims were made that bees failed all kinds of surprisingly simple tasks—discriminating squares from triangles, for example—had been flawed: since bees need to invest time scanning the shapes, a quicker solution for them was simply to be indiscriminate. Once penalties were introduced, bees suddenly did ten times better than they had in any previous experiment on color discrimination—but for difficult tasks, bees had to invest extra time in examining the stimuli on offer.

In previous decades, in virtually all work on animal intelligence, accuracy was the key parameter that was evaluated. Now it became clear that time was of the essence, too. An animal can place emphasis on either speed or accuracy, depending on context—sacrificing accuracy for speed if error is cost-free, but if error carries penalties, taking more time and care to choose correctly. Such speed-accuracy tradeoffs have since been confirmed in the decision making of many animal species other than bees. Associative learning, in which animals associate a stimulus with reward, is far from the simple process it was once believed to be. In bees and other animals, it involves attention and other cognitive processes.

## A Strange Flower in the Shape of a Human Face

A remarkable string of discoveries in recent years has revealed bees' tremendous flexibility in learning about almost *any* sensory input. Adrian Dyer discovered, for example, that honey bees could be trained to recognize images of human faces, even though human faces normally play little role in bees' lives. There was, at that time—also in the early 2000s—a debate among psychologists studying humans over whether face recognition is a specialized ability that requires a special face-recognition module in the brain. This is often assumed, but it is also possible, in humans and other animals, that it is simply a highly practiced skill, refined through repeated experience. If so, face recognition should also be possible with a non-specialist pattern recognition system, such as bees use for flower recognition.

Adrian and team used a series of black and white photographs of faces from a standard test used to diagnose a condition called prosopagnosia ("face blindness"), where human patients fail to recognize familiar faces. Bees were trained to associate a sugar reward with one face, while other faces were associated with bitter quinine solution. The bees performed remarkably well at a task failed by patients with damage to the so-called fusiform face area of the brain. The results thus showed that face recognition is technically feasible without specialized circuits—clearly the bee has no brain module for remembering human faces, so it is presumably using the same brain circuits it uses for flower recognition.

Nonetheless, in some social species where face recognition is important, such as humans, or some species of wasps that actually recognize nestmates' faces (see "Different Lifestyles, Similar Brains" in chapter 9), the brain circuits that mediate general pattern recognition may have been further tweaked over evolutionary time, turning such species into natural-born expert face recognizers. To our honey bees, the images of human faces associated with sugar rewards were presumably just "strange flowers," but this discovery certainly illustrated the versatility and flexibility of their learning ability.

## Learning about Flower Texture

Two British scientists who made many further discoveries on bees' flexibility in learning flower traits are Beverley Glover and Heather Whitney, both botanists rather than zoologists or sensory physiologists. Often, it is possible to deduce features of an animal's sensory world by exploring what is significant to these animals in their natural environment—in the bees' case, flowers. In 2004, I was approached by Beverley, who was interested in floral evolution. She had a unique study system—a set of mutant snapdragon flowers that differed from wild-type flowers in various subtle ways. A single mutation in these plants causes the flower petals to have a different texture and a different visual appearance. What might be the function of the natural texture for interactions with pollinators, and are these interactions disrupted in mutants?

Beverley's snapdragon flower mutants differ in one crucial aspect from wild-type flowers. In wild snapdragon flowers (and indeed in many other flowers), the epidermis (the flowers' "skin") consists of cone-shaped cells, whereas the cells of the leaves, for instance, are flat (figure 7.4). The mutants, instead, have the same flat cell shape in the flowers as in the vegetative parts of the plant. With postdoc Heather Whitney, we found that one function of the "rough" surfaces of flowers (generated by lots of microscopic cones side by side) is to assist bees to grip flowers with their feet when they land and to manipulate floral structures more easily for their rewards. But the cone-shaped cell protrusions also act as lenses that direct light into the pigment-filled compartments of the cells, which gives the natural flowers a richer, more saturated color—and it also results in the flowers being a few degrees warmer. Could this difference in temperature be of relevance for bee pollinators?

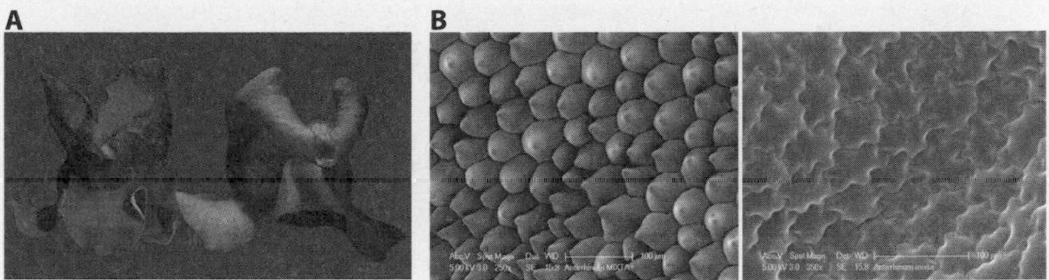


**Figure 7.4. Wild-type and mutant flowers of snapdragon, and their epidermis (flower surface) structure.** ***A.*** Wild-type snapdragon (*Antirrhinum majus*) flowers have a deeply saturated color (***left***), while so-called "mixta" mutant flowers appear more flat (***right***), despite having identical pigments. ***B.*** The only direct effect of the mutation is a change of shape in the epidermal cells, which are conical in wild-type flowers (***left***) and flat in the mixta mutant (***right***). As well as affecting bees' grip on flower surfaces, this has indirect effects on floral temperature and color, because conical cells act as lenses to focus the light into the pigment-containing vacuoles.

## How a Warm-Blooded Insect Learns about Floral Warmth

Contrary to the common belief that only mammals and birds are warm-blooded, bees are, too (at least a lot of the time). A bee's thorax temperature needs to be a minimum of 30°C for the bee to fly, and a bee's normal body temperature during flight can be ~40°C—at times a full 30°C above the ambient temperature in the case of the relatively cold-tolerant bumble bee. At cold ambient temperatures, bees can warm up to this level by shivering their flight muscles (figure 7.5). The carbohydrates contained in flower nectar are thus in part needed to keep the flight engine running—but an easy shortcut to obtaining the necessary warmth might be to seek out flowers that contain warmer nectar.

**Figure 7.5. White-hot pollinator.** Thermographic image of a worker bumble bee taken with an infrared camera: shades from blue through red and yellow to white indicate increasing temperatures. The thorax is hottest at ~38°C.

Our combined team of color vision scientists and botanists found that bees invariably preferred flowers with warmer nectar (even when the difference was just 4°C), a bit like us drinking a hot drink on a cold day. (If you need to warm up, you can produce your own heat, at some cost to your energy reserves—or you can consume a warm drink, and save your energy.) But bees could not sense the flowers' warmth from a distance. Instead, they had to sample the flowers' temperature first, and then they learned the associated colors of the flowers, enabling them to predict the temperature. Bumble bees came to prefer the flower colors that reliably offered them a warm drink. Such a three-way interaction between two types of rewarding stimuli (nectar and warmth) that needed to be sampled in direct physical contact and an associated longer-distance, visual signal (color) had not been described before.

## Bedazzled by Iridescent Flowers

The flexibility of bees in associating sensory cues with reward appears virtually limitless. It seems that if there is an appreciable sensory signal, bees can learn it. Heather Whitney and Beverley Glover had observed that some flowers generate a subtly iridescent effect—so that when viewed from different angles, the same areas on the petals might appear in different colors. But if flower color is meant to be a reliable, species-specific identifier, then wouldn't such iridescence (which can potentially include all colors of the rainbow) be highly confusing for bees?

Working with experts from the Cambridge physics department, the team discovered that the nanostructures of such flower surfaces took the form of diffraction gratings (tiny parallel ridges spaced 1–30 μm apart). Artificial flowers were then made by first making dental wax casts of real flowers that were subtly iridescent, and then imprinting these on epoxy resin. More strongly iridescent artificial flowers were generated by using diffraction grating film as a template structure for the dental wax casts. This made it possible to generate synthetic flower disks that combined various pigments (added to the resin) with various degrees of iridescence. The findings showed that bumble bees, far from being confused by the changing appearance of the flowers, actually used the very changeability as a cue, and could tell iridescent from equally pigmented non-iridescent flowers. The iridescence also made it easier for bees to locate the flowers in the first place.

Using more of her elegant artificial flower designs, Heather continued to explore the limits of the flexibility with which bees could integrate multiple sensory cues to identify flowers. For example, the bees' sensitivity to polarized light, discovered by Karl von Frisch, was traditionally thought to be used only for navigational purposes (to augment the sun compass)—but Heather, with collaborators in Bristol, discovered that bumble bees can actually learn polarization patterns presented to them on floral targets. However, this works only if the bees can approach the flowers from underneath, looking with their dorsal eye region—because it is only the dorsal area of the eye (which usually looks at the sky) that is polarization-sensitive. This clearly shows that many types of sensory input are not simply wired into particular behavior routines in an inflexible manner. Instead, if you reliably link *any* measurable sensory input to reward, bees will readily learn this connection (although the degree of their readiness to do so might vary with how likely that link is to occur in nature).

## How Bees Learn Rules

While I was still working on my PhD in Berlin in the early 1990s, a young Argentinean scientist, Martin Giurfa, joined Randolf Menzel's team as a postdoctoral scientist. We initially began collaborating on honey bees' innate color preferences, but did not get much further than confirming John Lubbock's nineteenth-century observation that honey bees are partial to the color blue. Soon after,

Martin started producing a string of now classic studies on the cognitive abilities of honey bees, first jointly with Randolf Menzel and later at the University of Toulouse, where he was soon to found his own research center.

These studies pioneered the question of how bees' minds categorize distinct objects by common features and form rules and concepts that link separate experiences over their lifetime. Martin was interested in whether honey bees could classify visual patterns by whether they are symmetric or asymmetric. Every single bee of the "symmetric group" was rewarded on a variety of completely different black and white patterns that had only one thing in common: they were bilaterally symmetric (could be mirrored along a vertical midline). During this training, bees also encountered a variety of asymmetric patterns that were *never* linked to reward. Now, bees could in theory solve this task by memorizing every single rewarded pattern presented during training—but this possibility can be probed in a so-called transfer test, when subjects are confronted with patterns they've never seen before that share the target feature of the patterns they were trained on.

The experiment demonstrated that bees had indeed formed a category of symmetrical patterns, all of which they classified as rewarding while shunning asymmetrical targets. Conversely, bees that had been rewarded on asymmetrical patterns during training preferred novel asymmetric objects in transfer tests. In small-brained animals, learning the common features of multiple rewarding patterns (such as those presented by flowers) might actually be a strategy to economize on memory storage space: rather than memorizing each rewarding pattern individually, it would be more efficient to memorize just the crucial features that link all of them.

With a team of international collaborators, Martin then explored the ability of honey bees to learn sameness-difference rules,

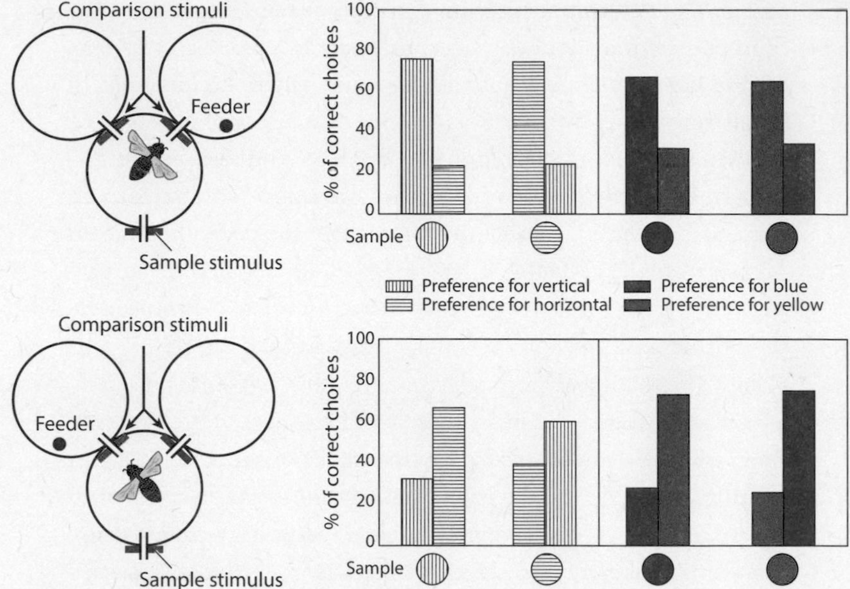


**Figure 7.6. Learning the concepts of sameness and difference in the honey bee.** ***Top row:*** Sameness rule learning: bees were trained in a delayed matching-to-sample paradigm, where they were first shown a sample of a visual stimulus (e.g., yellow) when entering a Y-maze type setup. At the decision point, they were presented with two stimuli (here yellow and blue), only one of which matched the one previously seen, and which marked the entrance to a chamber with a sugar reward. After training, bees will not only be able to reliably pick the same stimulus as seen earlier, but will learn a rule to "always pick the same one," no matter what the stimulus actually is, i.e., no matter whether it is yellow vs. blue or vertically striped vs. horizontally striped. ***Bottom row:*** Difference rule learning (delayed non-matching-to-sample): in the same apparatus, bees can also learn to "always pick the different one," so that they have to briefly remember the sample stimulus, and then choose the stimulus that differs from it.

where subjects are trained in a so-called "delayed matching to sample" task. When learning the concept of sameness, honey bees were first shown a sample visual pattern A, then given a choice between A and B, where A was associated with a reward (figure 7.6). Thus, bees needed to store the first pattern in *working memory*—the short-term, low-capacity kind of memory that, in humans, is used between being told someone's phone number and writing it down. Bees store the first pattern in working memory as they enter a maze; inside the maze, two patterns are on display, one of which is the same as the one seen earlier (the matching stimulus). After repeating this sequence with a series of varied stimuli, bees eventually learned a more general rule—to "choose the one that is the same"—even with entirely novel sets of stimuli (for example, pattern C followed by C and D) that they had never seen before. Honey bees could also learn the inverse concept: "choose the different one." Subjects learned always to choose the pattern that was *not* the same as the sample—the "mismatch."

As noted, these tasks involve so-called working memory—a transient form of memory that in bees and other animals lasts only a few seconds, and is more limited in capacity than long-term memory. In this case, the task was for the bee to interrogate the contents of its working memory ("What stimulus have I just seen?") and to compare the incoming stimuli with it; then, in the case of the "sameness task," to pick the target that matched the one in working memory, or in the "difference task," to pick what was *not* in the short-term memory buffer. Perhaps most impressively, bees could transfer the "sameness" and "difference" concepts between sensory modalities—applying them to visual stimuli, for example, even when they had been trained on olfactory stimuli.

The implications of this study, showing that an animal as small-brained as a bee could learn rules, were so wide-ranging that it caught the attention of Christof Koch, an American neuroscientist renowned for exploring the neural basis of consciousness in humans. In an article in *Scientific American*, Koch observed, "Bees display a remarkable range of talents—abilities that in a mammal such as a dog we would associate with consciousness." We will return to the question of consciousness in more detail in chapter 11. For now, another factor that matters in the context of flower visitation (and of consciousness) is learning about *time*.

## Learning *When* Flowers Are Rewarding

It has been known since Buttel-Reepen's investigations in the late nineteenth century (see chapter 4) that bees can remember the time of day at which different flower species produce nectar. Martin Lindauer discovered that dancing honey bees can even retrieve such memories "off-line" in the depth of the night (see chapter 6). However, the complexities of the flower supermarket mean that bees have to keep track of multiple contingencies over several different timescales. While flowers of a species may secrete nectar only at certain times of day, the nectar also refills after it has been emptied by a flower visitor, sometimes in less than an hour, though at different rates in different flower species.

The Canadian scientists Michael Boisvert and David Sherry directly measured whether bees could learn the interval between two flower rewards, and could accordingly *wait*—i.e., withhold sticking out their tongue—in the meantime. Indeed, such self-control is often regarded as an indicator of superior intelligence in large-brained vertebrates, including humans. The scientists trained groups of bumble bees to be rewarded at intervals of either 6, 12, 24, or 36 seconds—and sure enough, the bees of each group managed to keep their responses in check until the trained interval was over. Thus, in essence, Boisvert and Sherry's bees were predicting the future, in that their behavior suggests an expectation tuned to the lapse of different time intervals. It has been suggested that this ability comes in handy when bumble bees plan routes for visiting multiple flowers with familiar nectar replenishment rates, and that this familiarity gives seasoned foragers an edge over casual visitors to a flower patch.

## Learning Spatial Concepts

A student of Martin Giurfa, Aurore Avarguès-Weber, discovered that honey bees can solve a task that in primates is viewed as spatial concept learning: the concept of "above" and "below." Two pairs of visual stimuli were shown in the two arms of a Y-maze. Each arm displayed one pair: a "referent" shape (e.g., a horizontal line), and a second geometric shape that appeared either above or below the referent (figure 7.7). Bees learning the "concept of aboveness" had to choose the arm of the Y-maze in which a shape—any shape—occurred above the referent, while those learning the "concept of belowness" had to pick the arm in which there was an arbitrary item beneath the referent. Bees swiftly learned this task, and were even able to pass a transfer test when presented with a completely unfamiliar target, which, however, appeared again in the correct location relative to a familiar referent. Aurore observed that bees were quicker at learning these tasks than primates, which take hundreds or even thousands of trials to master similar tasks; the bees took only a few dozen. Human infants, with their perhaps 100 billion neurons, only pass such tests by the time they are six months old. We will find out in the next section, however, that honey bees solve the

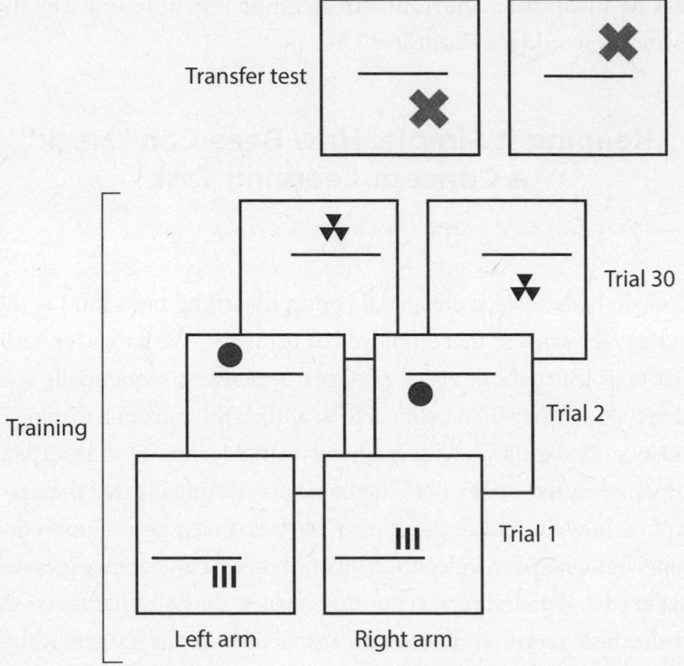


**Figure 7.7. Training schedule for honey bees learning the concept of "aboveness."** Shown are pairs of stimuli sequentially presented to bees on the back wall of a Y-maze. Bees were trained to collect sucrose solution in one of the arms of the maze. Each pair of patterns simultaneously presented in the Y-maze contained the same visual stimuli, only distinguished by whether the "target" occurred below or above the "referent," which was always the same (horizontal bar). Bees encountered a series of different targets during training, and were rewarded only if they chose the pairing that contained the target above the referent (another group of bees were rewarded if they correctly picked the arm where targets consistently appeared below the referent). Please take a moment to solve the task for yourself—it helps to understand the challenge with which the bees were faced. The exact positions of targets and referents on the back walls of the arms of the Y-maze were varied to ensure that bees could not learn a fixed configuration. (The correct choices are: right—left—left—right.)

task by means of a shortcut—in a completely different way than primates would go about it.

## Keeping it Simple: How Bees Can "Hack" a Concept Learning Task

How do bees solve such spatial concept learning tasks, and is their strategy the same as that employed by primates? We have seen earlier that bees learn about visual patterns by actively, sequentially scanning them, rather than being able to absorb them from a distance at a glance. Two PhD students from my team, Marie Guiraud and Mark Roper, analyzed honey bees' flight trajectories inside the Y-maze to explore how they solve the aboveness / belowness task. Hundreds of hours of high-speed video footage of the bees' choice strategies were analyzed in detail. It turned out that most of the bees that succeeded in the task required a close-up inspection of the targets—but it turned out they only looked at the bottom part of the pattern before making the decision to land on a feeder. How could they have solved this task by only looking at one of two shapes in each pair? It's important to explore this in detail, since it shows that animals come up with wholly different strategies to solve a cognitive task than humans might—and these strategies can nonetheless be exceptionally nifty.

The tested bees often had a side bias (favored either the right or left arm of the Y-maze) or chose an arm randomly. To examine what happened next, let's focus, for the moment, on bees trained to the "aboveness" task. Upon arriving at the back wall of the maze showing target and referent (figure 7.7), they flew to whatever happened to be the bottom item. If this was the referent (the horizontal line in figure 7.7), the bees knew they had arrived in the correct place: no further inspection of the item above the referent was necessary. Any arrangement with the referent at the bottom is the correct one. If the lower item was *not* the referent, the bee again did not have to inspect any other item in the arrangement—she knew that she was in the wrong arm of the Y-maze, and simply had to fly into the alternative arm to collect her reward.

Bees in the "belowness" tasks used the reverse strategy. If the Y-maze arm initially chosen contained the referent at the bottom—wrong place; fly to the other arm (no further inspection of any other item necessary). If, in the arm initially chosen, *anything but the referent* was the bottom item, the bee was in the right place. This means that honey bees had essentially solved what scientists *thought* was a concept learning task—by a strategy that did not require them to learn a concept at all. Of course, this doesn't mean that they couldn't also potentially solve it in the way that humans, or at least experimental psychologists, would expect—and some bees might do so with more-extensive training.

In this chapter, we have explored bees' psychology in the context of flower visitation. It is clear that bees do much more than "store an image" of a flower type after they have found a reward inside. What is stored is a rich sensory experience, making use of input from all sense organs available—and experiences from multiple flower types can be merged and common features extracted. We have learned that the intelligence of bees is superb, but different from that of other animals, including humans. No matter the particular solutions that they have come up with, it is clear that bees display cognitive capacities that until recently had been considered the domain of much larger-brained vertebrates. This might be surprising in light of the widespread view that large brains are required for, or at least somehow associated with, intelligent behavior. But an inspection of the literature perpetuating this view reveals that scientists often rank cognitive abilities of animals from "simple" to "advanced" based on intuition, rather than on analysis of the neural computations actually required for such abilities. We will turn to this question in chapter 9, on the bee brain, where we will discover that many apparently difficult learning tasks can in fact be mediated by very modest neuron numbers.

Here, we have already learned that seemingly basic processes such as "simple associative learning" (as in the pairing of color with reward) are far from simple, and that performance depends on context, motivation, and attention. On the other hand, bees (and, in principle, other animals) can solve what appears to be a complex "concept" learning task using relatively simple tricks. In the same vein, category formation (sorting multiple individual stimuli by common identifying features, such as symmetric versus asymmetric flowers) might be viewed by many as more cognitively advanced than memorizing individual flower patterns. That might well be the case—unless storage space is constrained. In small animals such as insects, which lack carrying capacity for a larger brain, the physical limits on memory might actually impose a selection pressure for clever modes of information storage, such as by category formation. Taken together, these findings do not just call into question the notion that large brains are required for any cognitive capacity. The same cognitive capacity might be mediated by entirely different strategies in different species, and the neural circuits and computations underpinning them might therefore also be different. In fact, before we understand a cognitive operation as a neural circuit function, we should be cautious about classifying it as a "lower" or "higher" form of cognition.

In the preceding chapters, we have explored the learning abilities of bees as individuals. Yet the species we have mostly examined are social, and in the coming chapter we will investigate the many remarkable ways in which social bees can learn from each other, including unique symbolic communication systems, a form of democratic decision making, and the ability to copy one another in simple "tool use" tasks.

# 8

# From Social Learning to "Swarm Intelligence"

> In the summer of 1857 I observed a much more curious case of one insect apparently imitating a complex action from another of a different genus . . . I saw several humble-Bees . . . cutting with their mandibles through the underside of the calyx [of kidney-bean flowers], and thus sucking the nectar. . . . The very next day I found all the hive-bees without exception sucking through the holes, which the humble-bees had made. . . . I must think the hive-bees either saw the humble-bees . . . and understood what they were doing . . . ; or that they merely imitated the humble-bees. . . . .
>
> Should this be verified, it will, I think, be a very instructive case of acquired knowledge in insects. We should be astonished did one genus of monkeys adopt [a feeding behavior] from another . . . ; how much more so ought we to be in a tribe of insects so pre-eminent for their instinctive faculties, which are generally supposed to be in inverse ratio to the intellectual!

**—Charles Darwin, 1884 and 1841**

Social learning—the ability to learn by observing others (typically but not always of the same species), or under the influence or guidance of others—is considered one of the basic building blocks of human culture. Does this mean that such learning is unlikely to exist in small-brained animals such as bees? We have already learned about a specialized form of social learning, the honey bee dance, in chapter 5. But can bees also learn from each other by observing other bees' behavior in the field, as Darwin suggested?

## Learning from Other Bees Which Flowers to Visit

There are good reasons to explore social learning in pollinating insects, which must compare the nectar and pollen offerings of plant species and seek out the best bargains. Gaining reliable information on what constitutes a good flower species or patch often requires extensive sampling, and as we have seen, in a species with complex flower morphology, the particular manipulation techniques to gain access to nectar can take several dozen trials to master. Since many pollinators (typically of multiple species) often work concurrently in a meadow, there is ample opportunity to pick up information from others. Moreover, social insects live in colonies containing dozens to thousands of closely related individuals. These "superorganisms" have a need for active information sharing and learning from each other that may be unparalleled in the vertebrate world.

In any inexperienced bee's flight range there will often be multiple flower species, each differing in the quality and quantity of nutrition it offers. The rookie might either sample all these flower species, or alternatively, pay attention to the foraging activities of other pollinators. A useful shortcut to finding a profitable resource might be to begin foraging where other flower visitors are active—especially where these are of the same pollinator species, and thus share the same nutritional needs and physical fit with flowers. To explore whether bumble bees make use of this potentially important source of information, my PhD student Ellouise Leadbeater built a simple "meadow" with two equally rewarding artificial flower "species"—one with blue and one with yellow flowers. Naïve individuals (that had never visited flowers before) strongly preferred to land on inflorescences occupied by an experienced bee. After that, beginners typically stuck with the flower species they had first been guided to; if they switched at all, it was almost invariably when they could observe a conspecific on the alternative flower species. Clearly, then, these novices were not just attracted to fellow bumble bees under any circumstances; they only joined seasoned foragers when these held out the promise of a novel type of bounty.

## Learning by Observation of Other Bees from a Distance

In a parallel study to Ellouise Leadbeater's, Bradley Worden and Daniel Papaj at the University of Arizona, Tucson, also found that naïve bumble bees were not simply attracted to fellow bees. In their study, the novices were separated from the "demonstrator" bees by a glass screen. After having observed the seasoned foragers from a distance on one of two flower colors, the observers were placed on their own with the two flower types—and they strongly preferred the flower color that they had witnessed being visited earlier by their compatriots.

This finding was remarkable in that at the time of observation, no reward was given to the bees that were locked behind the glass screen—they could merely see the foraging bees' activities from afar. Nonetheless, they had learned the rewarding flower color. Did these observer bees "understand" what they were seeing, as Darwin suggested? We tried an explanation that at the time seemed more parsimonious—one that would involve classical associative learning, albeit in two steps. The process we suspected might be at work is called—since Pavlov's time—"second-order conditioning": an individual first learns that stimulus A is linked to reward and then sees stimulus A and B together, and as a result predicts that B is also rewarding.

Another PhD student, Erika Dawson, and Aurore Avarguès-Weber joined the team to explore whether bumble bees display such second-order conditioning when learning from each other how to identify the most rewarding flowers. They began by testing whether bees can learn to associate other bees (stimulus A) with the presence of reward. Subsequently, when the bees so trained see fellow bees on a particular flower color (stimulus B), they might "conclude" that since A is known as predicting reward, and A was seen with B, that B must also be rewarding (figure 8.1). Indeed, the experiment suggested not only this, but also that the meaning of seeing a conspecific on a flower could actually be reversed: if bees had come to associate the presence of other bumble bees with bitter quinine, they would subsequently pick a flower color they had *not* seen demonstrators visiting.

We already know from chapter 6 that bees can learn important information without rewards, but by observation, such as during orientation flights in which they acquire information about the landmarks surrounding their hive. Our findings above showed that bees can also associate two stimuli by viewing them from a distance, while only having background knowledge that one of the stimuli is

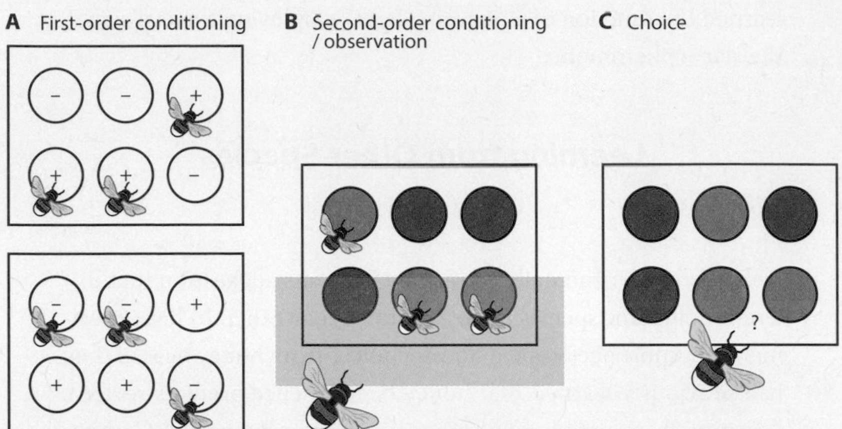


**Figure 8.1. Social learning by observation—explained by second-order conditioning in bumble bees?** A seemingly complex behavior (copying the flower choices of other bees) could emerge from associative learning, whereby foragers use the appearance of conspecifics as a predictor of sucrose reward (+ in panel ***A, top***) or bitter quinine solution (– in panel ***A, bottom***). When the observers so trained would later see conspecifics visiting green flowers through a glass screen (panel ***B***), they would link the bee presence to what they had learned about its significance earlier (panel ***A***). When the observers were then given a choice in the absence of demonstrators, they would prefer green if their first step of training had informed them that the presence of other bees means reward, or orange if they knew that conspecific presence signifies bitterness.

rewarding. The flexibility in making such associations was further underlined when Erika Dawson, working with Elli Leadbeater, discovered that honey bees could learn to react to an arbitrary stimulus (a colored light) as a threat indicator by pairing it with the alarm pheromone that is normally released when honey bees organize an attack on a predator invading the nest. The bees had essentially learned the function of a "warning light" employing a social signal, the alarm pheromone.

## Learning from Other Species

Erika also experimentally confirmed Darwin's suggestion that different pollinator species might "imitate" each other. In her experiments, bumble bees copied flower choices from honey bees, if they had previously learned that honey bee presence predicts reward. Learning from members of other species can be useful. Competition among conspecifics, especially if they are members of the same colony, can lead to overexploitation of the same resources and escalation of the competition. In this context, monitoring the food choices of individuals of other species that share similar food sources and habitats can be a rewarding strategy, alerting a forager to the location of novel, potentially more profitable resources. Other species may differ in their vigilance levels, perceptual capacities, or information-gathering methods. Thus, observing them can provide access to information that is hard to obtain by individual sampling or by observing conspecifics. Competition might also be less pronounced between species whose demands only partially overlap.

A study in which two different species of honey bees, *Apis mellifera* and *Apis cerana,* were housed together in the same hive revealed that bees can learn to decode the dance language (see chapter 5) of another bee species. The two species differ in their "distance code," so that the same food source location is indicated subtly differently by dancers of the two species. When "talking to each other" in the same hive, however, *A. cerana* can learn to decode the dance "dialect" of *A. mellifera,* presumably by a form of trial-and-error learning. The *A. cerana* foragers must first notice that their initial erroneous reading of the dances leads them to a reward-less location. When a subsequent search takes them to the rewarding site, their reading of the distance code apparently recalibrates, so that thereafter they read the "foreign dialect" correctly. Turning to stingless bees, *Trigona spinipes* bees can learn about the foraging sources of other bee species by engaging in espionage of the scent trails that those other bees lay to a rich food source. They then take over that food source, driving away or even killing the bees that had originally discovered and exploited the resource.

## Learning Tricks of Thievery by Observation

Darwin (see this chapter's introductory quote) suggested that bees could learn from each other the particular techniques of how to access rewards in flowers. Specifically, he described what is now called nectar robbing, the crafty technique some short-tongued bees use to exploit long-tubed flowers: biting a hole in the calyx, "stealing" the nectar without necessarily pollinating the flowers.

Elli Leadbeater presented long-tubed flowers that could be exploited either "legitimately" (by bees crawling into the tubes) or by bees creating orifices at the base of the tube, using their mandibles (figure 8.2). Under such conditions, most bumble bees by default took the "long route" to the nectar; only a minority of bees "innovated" to puncture the nectar spurs spontaneously. However, after some bees had bitten holes into the base of the tube, more and more bees switched to nectar robbing, initially by using the holes created by other bees. Once bees had experienced this shortcut to the goodies, they became more likely to bite holes into the base of the flower themselves. Such social learning thus generates an

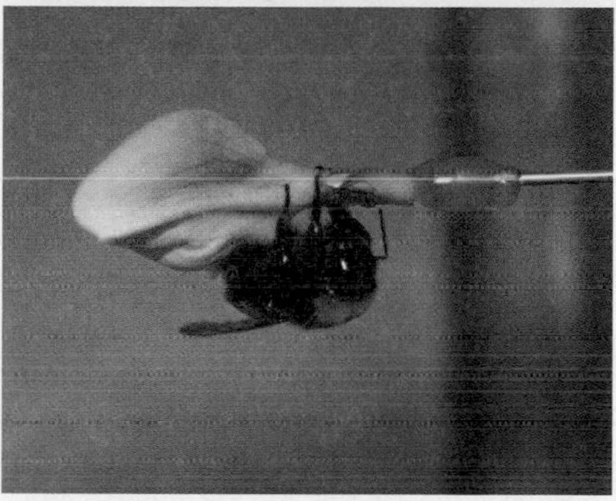

**Figure 8.2. Nectar robbing by bumble bees.** The images show bean flowers connected to a nectar pump. ***Top:*** A bumble bee forager enters the flower "legitimately," extracting the nectar while touching the flower's reproductive parts. ***Bottom:*** A "nectar robber" extracts the reward through a hole bitten into the base of the flower, without providing pollination service. Naïve bumble bees can learn this technique by observing experienced nectar robbers.

opportunity to explore how group-specific behaviors spread from "innovators" to others in the group.

Dave Goulson, a British conservation biologist, observed the nectar-robbing habits of Alpine bumble bees on a wildflower called the yellow rattle (so named after its dry fruit, which contains loose seeds). Its long-tubed flowers have two vulnerable areas (one on each side) close to the nectaries, and short-tongued bees are quick to figure this out and start biting holes into either the right or left side of the flowers. Goulson and colleagues discovered that there were apparently local "traditions" of right-handedness or left-handedness: patches of flowers tended to be exploited either one way or the other, and patch-specific biases increased further throughout the blooming season. This suggests that the initiators of nectar robbing in each patch might swiftly specialize on either the right-handed or left-handed approach. Such specialization in turn creates opportunities for observers to either copy the technique, or to discover and use the holes bitten by the early innovators.

## Culture and Tradition in Bees?

We bumble bee researchers were not the first to suggest that behavioral traditions might exist in bees. In the early 1980s, Martin Lindauer (chapter 3) explored the elements of cultural traditions in honey bees. In one of his final experimental studies, he explored how the daily rhythm of a honey bee colony's activities, shaped in response to daily patterns of food availability, is then transmitted as "traditions," even to new foragers who have never directly experienced those environmental conditions.

Lindauer taught groups of foragers that food was available for only one hour a day, either 5:00–6:00 in the morning or 8:00–9:00 in the evening. These bees learned to match their activity pattern to the availability of food. Brood cells were then removed from each colony, and larvae and pupae developed into adults in an incubator, without contact with their older nestmates. After hatching, these bees' preferred activity patterns matched the highly unusual ones of their mother colonies—they were either "early risers" or "late-night workers." The exact mechanism by which the young brood becomes conditioned to the colony's activity period remains a mystery; Lindauer suspected that increased vibrations of the comb caused by dancing bees during periods of heavy recruitment may be picked up by the larval bees or pupae inside their comb cells.

The forms of social learning explored in the preceding paragraphs center on foraging phenomena that are relatively close to a bee's natural biology, in that they mostly focus on tasks of evaluation and manipulation encountered daily in flower visitors' lives. One of the hallmarks of human cultural traditions, however, is that they include phenomena demonstrably remote from anything to which our species might be innately predisposed. Would it be possible to monitor the spread of such a non-natural behavior in bees—something they don't typically display in nature?

In 2008, I founded a psychology research center at Queen Mary University of London, and was happy to employ several scientists exploring the cognition of the reputedly smartest of the nonhuman vertebrates, such as corvid birds and chimpanzees. In the study of vertebrate intelligence, it is common to confront animals with challenges that lie outside their natural ones—such as, for example, the string-pulling puzzle, in which animals need to grasp a string and pull on it to gain access to an otherwise unobtainable reward. I still remember the meeting when one of the bird researchers was explaining how some experimental parrots had failed this task. I exclaimed, "I bet our bumble bees can do it!"

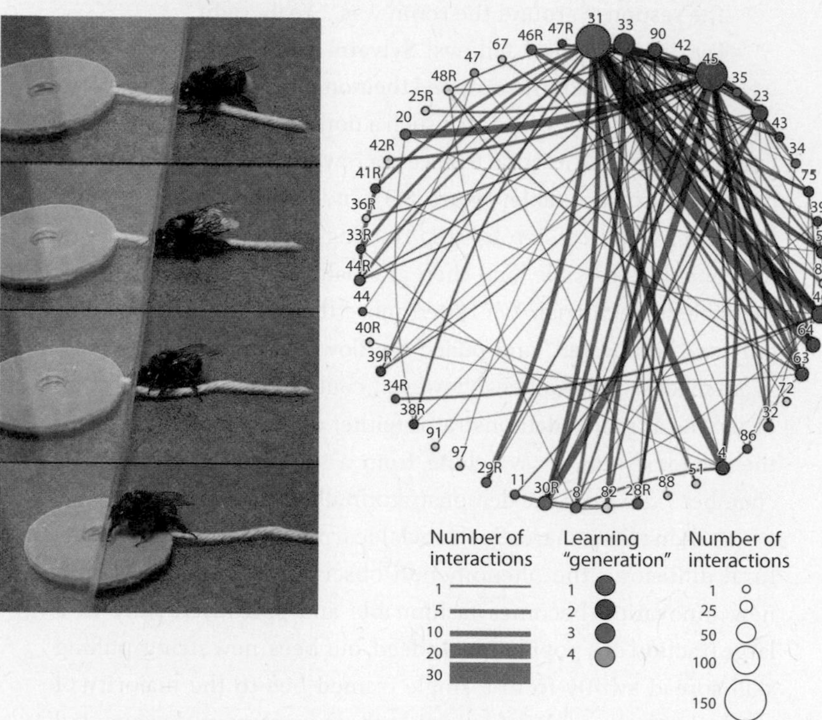


**Figure 8.3. String pulling by bumble bees.** ***Left:*** The image series shows a bumble bee forager pulling a string to gain access to a blue artificial flower under a transparent plexiglass table; the center of the flower holds a droplet of sucrose solution. ***Right:*** Diffusion of string pulling in a bumble bee colony. Numbered nodes designate individual bees. Lines indicate that two bees interacted at least once. Thickness of lines represents the number of interactions between two individuals. Size of nodes indicates number of interactions of that individual bee with any other bee. Color represents the learning "generation" of that bee. The first skilled bee is marked yellow and at the top position. Once a bee learns to string-pull, its node is orange for a first-order learner (learning from the "yellow bee"); pink for a second-order learner (interacting with first-order and lower-order bees); turquoise for a third-order learner (interacting with second-order and lower-order bees). Gray nodes mark bees that did not learn the technique over the observation period.

The response around the room was, "Yeah, right."

Two postdoctoral fellows, Sylvain Alem and Cwyn Solvi, weren't discouraged, and dropped their ongoing projects to explore whether bumble bees can learn such a non-natural object manipulation task: pulling a string to access a reward in an artificial flower under a plexiglass table (figure 8.3). Indeed they could, though most bees required step-by-step training where the "flower" was first openly accessible, and then gradually pushed further under the table at each trial. We tested more than 100 individuals, but only two "innovated" and pulled the flower from under the table spontaneously. Naïve bees, however, could also learn the task by observing a trained demonstrator (either directly interacting with the demonstrator, or watching from a see-through observation chamber 7 cm from the demonstrator pulling the string).

It didn't stop there. Such social learning led to veritable cultural diffusion—the phenomenon observed in humans when a new innovation becomes fashionable and spreads rapidly to a large fraction of a population. Indeed, our bees' new string-pulling skill spread swiftly from a single trained bee to the majority of a colony's foragers. We found that there were several sequential sets ("generations") of learners: previously naïve observers could first acquire the technique by interacting with skilled individuals and, subsequently, themselves become demonstrators for the next "generation" of learners. In this way, the longevity of the skill in the population could outlast the lives of informed foragers (figure 8.3).

At the time of our string-pulling work with bumble bees, we suggested that there was nothing particularly ingenious about the bees' performance in this experiment. We thought that the results were explicable by a combination of attraction to conspecifics, associative learning (learning that the presence of conspecifics signifies reward), and trial-and-error learning (to figure out the actual string-pulling technique), and that this combination might suffice for the cultural spread of foraging techniques. We had no clear indication that bees watching other bees solve the task "understood what they were doing" (as Darwin had put it), or that they attempted to imitate the behavior they observed. Perhaps there is nothing special at all about social learning, or about the presence of a conspecific in a learning setting; perhaps a fellow bee is just like any other item in a bee's environment that can be linked flexibly to a reward or a threat . . . ?

It's not that simple. To any animal, conspecifics have a distinct salience: their recognition is crucial even for solitary species, in selecting partners for reproduction. This applies all the more to social species that depend on one another for shelter building, provisioning, and mutual defense. And it turns out that even though bees can learn from members of other species, conspecifics are more "convincing" as sources of information about which resources to exploit. This makes sense, since members of your own species will typically have the same nutritional requirements as well as the same morphological adaptations for harvesting nectar or pollen (tongues of a particular length, mandibles of a certain strength, etc.).

But the special nature of conspecifics in social learning becomes obvious when they are replaced in an experiment with lifeless cues, such as wooden blocks, that have been set up to carry the same information as indicators of flower reward quality. In such cases, it is clear that novice bees lend more credence to live bees on the flowers than to model bees or any other objects. Using a device built from my childhood Fischertechnik construction toys to move model bumble bees on artificial flowers, we found that the movements of live demonstrators are crucial in directing the attention of observers to the flowers of interest.

## Learning Tool Use by Observation

In one set of experiments, we explored Darwin's suggestion that bees might not merely copy what they observed others to do, but that they also had some form of understanding of the desired task outcome. For this study, Cwyn Solvi teamed up with Finnish postdoctoral fellow Olli Loukola. In their study, bumble bees had to move a ball to the center of a circular arena to gain access to a reward, a task equivalent to token or tool use. Bees that observed demonstration of the technique by a live or artificial demonstrator learned the task more efficiently than did bees observing a "ghost" demonstration (ball moved via a magnet underneath the platform) or no demonstration.

To see if observers understood the desired outcome of the task, we played a simple trick on the trained demonstrators. We placed three balls in the arena at different distances from the target location. The obvious best solution to solve the task would be to move the closest ball to the center (figure 8.4). However, the two closest balls had been glued to the floor, and so the demonstrators had learned before they were paired with an observer that they could only move the farthest ball. And so this was the scene that the observers viewed when they were let into the arena with a trained demonstrator. Three times, the observer was present when the demonstrator moved the farthest available ball to the goal. After this they both received a generous reward of sucrose solution—sufficient for both bees to fill their crop.

Note that the observer herself had no direct experience of ball-rolling—she merely witnessed the procedure three times. After this exposure, the observer bees were put to the test alone (but all three balls could roll freely). In most trials, bees spontaneously

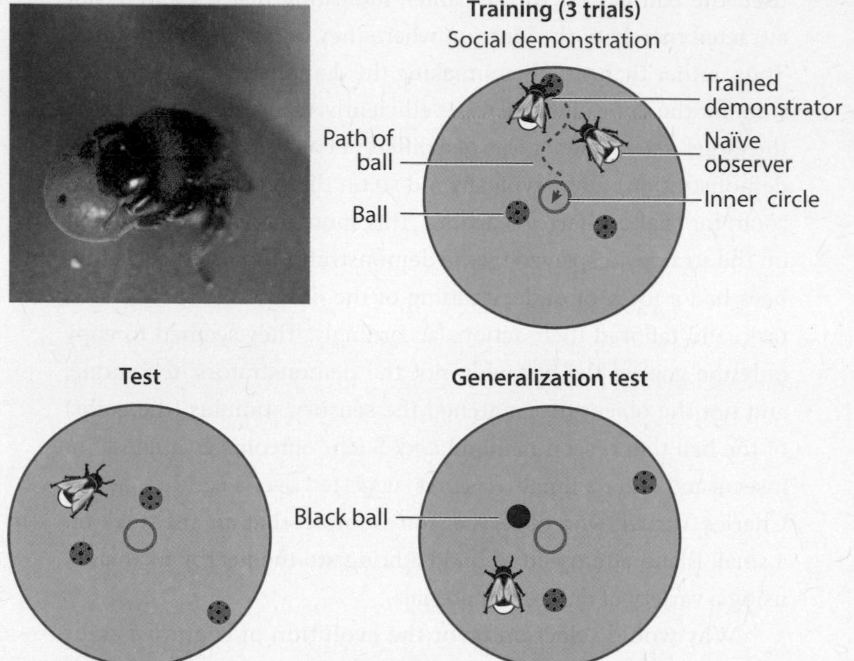


**Figure 8.4. Social learning of token use in bumble bees.** ***Top left:*** Bumble bee worker moves a ball by walking backward, holding on to the ball with its front legs. ***Top right:*** The task is to move a yellow ball to the center of the round blue area (the inner circle marked in yellow). An experienced worker (demonstrator) has learned that only the farthest of three balls can be moved (the others are glued down). A naïve observer watches. ***Bottom left:*** The observer is subsequently given the choice between three mobile balls. Rather than choosing the farthest ball (the one it has seen the demonstrator move), it picks the one closest to the center, a superior version of solving the same task. ***Bottom right:*** Even if the bee is faced with a novel black ball, it still chooses the optimal solution of picking the closest ball, rather than adjusting the arrangement to the one it is used to getting a reward with (yellow in middle).

used the ball closest to the center, indicating that they were not attracted merely to the location where they had seen conspecifics. Thus, rather than simply mimicking the demonstrators, observers went for the same end goal more efficiently, using the ball closest to the target (even when it was of a different color than the one in the demonstration). They typically did so on the first trial, leaving no room for trial-and-error learning. This spontaneous improvement on the strategy displayed by the demonstrator indicated to us that bees had a form of understanding of the desired outcome of the task, and tailored their actions accordingly. They seemed to copy only the goal of the behavior, not the demonstrators' technique, and not the object that matched the sensory stimulus (the color) of the ball they'd seen manipulated. Such "outcome awareness" in insects and other animals was first suggested over a century ago by Charles Turner, who observed, for example, that an ant stuck on a small island attempted to build a bridge to the nearby mainland, using a variety of different materials.

Why would selection favor the evolution of cognitive skills that are rarely or never encountered in nature, such as understanding where to place a ball to obtain food? The advantage of a form of general intelligence, which allows flexible problem solving, is that it allows coping with unpredictable challenges. Following the publication of the ball-rolling study, a member of the public emailed to report that she had observed a bumble bee rolling a small slug out of the nest entrance into which the slug had strayed, using the same technique as the bees in our study. Such rare challenges can be crucial to survival if access to one's young or one's food storage is at stake.

We have seen throughout this book that innate predispositions and learning go hand in hand. The inherent interest in conspecifics draws attention to their strategy in exploiting food sources—which in turn facilitates learning from them, in ways that are not strict copies of the observed behavior, but even sometimes spontaneous improvements. And while a same-species bee may be the most compelling tutor for a bumble bee, it is not the only acceptable one. Again we find flexibility: if members of other species reliably indicate reward, they too will come to be acceptable role models.

## Swarming to a New Home

Fascinating interactions between innate behavioral routines and social learning also take place in the context of bee communication. We have already seen an example in the honey bees' dance communication. While the symbolic codes for distance and direction are largely innate, the information is learned by "dance followers" inside the nest and subsequently applied when they are flying outside it. Another scenario of interplay between innate and learned communication behaviors is when honey bee colonies fission. When this happens, a substantial fraction of the colony's workforce (some 10,000 bees) leaves the hive or native nest, taking the old queen with them, and searches for a suitable new home. In preparation, the queen is constantly "harried" (bitten, pushed, or vibrated) by worker bees to slim her down for swarming (after she has been a cave dweller since at least the previous summer). Before it's time to go, active workers also harass other, still-sedentary workers by shaking them and finally by "buzz runs," plowing through crowds of inactive workers to rustle them up.

The actual swarming behavior has been described in rather poetic terms by the Nobel laureate in literature Maurice Maeterlinck (1862–1949), whose thoughts on the challenges of understanding bees we read in the epigraph to the introduction. Wealthy and a tad eccentric, Maeterlinck resided in an abandoned abbey with his (married) girlfriend, the author, actress, and singer Georgette Leblanc (1869–1941), and moved through the abbey's vast interiors on roller skates. Like many of the world's finest weirdos, he was fascinated by bees, and in 1901 he penned a magnificent book about them entitled *The Life of the Bee*. Of the swarming honey bees, he wrote: "The bees have already given the signal for departure . . . it is as though one sudden mad impulse had simultaneously flung open wide every single gate in the city; and the black throng . . . pours forth . . . in a tense, direct, vibrating, uninterrupted stream that at once dissolves and melts into space, where the myriad transparent, furious wings weave a tissue throbbing with sound."

If you think Maeterlinck got a little carried away here, see it for yourself one day. My former postdoctoral fellow James Makinson recalls that observing this spectacle just once made him want to become a biologist. Buttel-Reepen (1900) describes the mental state of the swarming bees as a sort of intoxication (*Schwarmdusel*) that comes with reduced aggression and increased attraction to light, and he muses whether it could involve a form of pleasure, like that inherent in play behavior. There are many mysteries involved in how these swarms are initiated, and indeed how it is decided who leaves and who stays at home. It is also remarkable that bees would decide to leave behind the fruits of many months of their labor (the wax comb construction, the larvae brooding in it, and their honey and pollen stores) and fly into a wholly unknown future, where even in the best of conditions, having to start from scratch is a certainty.

What happens after the beard-shaped swarm has settled on a tree branch is well investigated. The bees must find a suitable new accommodation. The stakes are high. Western honey bees naturally nest in tree cavities—but not any cavity will do: it has to be large enough but not too large, not too humid, have a reasonably small opening that can be defended and protected against inclement weather, and so on. A key challenge in moving to such a site is not just finding it, but also getting the swarm to agree on a single location. Disagreement is not an option. Workers can't survive on their own for long, and not even in large groups without the queen, so they have to reach a consensus before they run out of food or weather conditions become a significant risk. The new home must ideally be located in the midst of rich flower foraging opportunities, since the bees must fill the new nesting space with honeycomb and resources sufficient to survive the winter. This is a substantial challenge, and bees that make a poor choice often starve or freeze to death in the cold season.

According to Maeterlinck, this is how the decision-making process might work:

> The swarm . . . will remain suspended on the branch until the return of the workers, who, acting as scouts, winged quartermasters, as it were, have at the very first moment of swarming sallied forth in all directions in search for a lodging. They return one by one, and render account of their mission, and as it is manifestly impossible for us to fathom the thought of the bees, we can only interpret in human fashion the spectacle that they present. We may regard it as probable, therefore, that most careful attention is given to the reports of the various scouts. One of them it may be, dwells on the advantage of some hollow tree it has seen; another is in favor of a crevice in a ruinous wall, of a cavity in a grotto, or an abandoned burrow. The assembly often will pause and deliberate until the following morning. Then at last the choice is made, and approved by all. At a given moment the entire mass stirs, disunites, sets in motion, and then, in one sustained and impetuous flight, that this time knows no obstacle, it will steer its straight course, over hedges and cornfields, over haystack and lake, over river and village, to its determined

> and always distant goal. It is rarely indeed that this second stage can be followed by man. The swarm returns to nature; and we lose track of its destiny.

It is astounding how Maeterlinck guessed largely correctly, given that little was actually known about the existence or activities of honey bee scouts, or the bees' ways of communicating. But it was already likely, from earlier reports, that the bees in the swarm had made their decision prior to departure from the interim cluster site. The somewhat verbosely named Baron August Sittich Eugen Heinrich von Berlepsch (1815–1877), more succinctly called the Bee Baron, described in 1852 how he watched a swarm that had settled on a lime tree on his estate. He observed how bees departed in all directions initially, making curvaceous orientation flights. The swarm was still in place the next morning, and the Bee Baron ordered his stable boy to prepare for the pursuit of the swarm to its destination by having two horses saddled and all the estate's gates opened so they could chase after it without delay when it departed.

At some point multiple bees were seen leaving in a southerly direction in a straight line, and shortly afterward the entire swarm lifted off, due south. The two riders followed the swarm, which moved at an altitude of "4–9 feet" with "the platoon leaders clearly visible at the tip of the swarm." For the first 15 minutes, the horses' medium-speed gear (trot) sufficed to keep pace with the swarm, but then the bees sped up so the riders had to switch to galloping. The Bee Baron reports that "in the next village, not quite 45 minutes away, the swarm descended in a farmer's garden. As if in a coursing hunt, I jumped over the fence, found myself in the center of the swarm with my horse, and saw how the bees moved into a hollow pear tree. This materialized with such a speed that I had no doubt whatsoever, that the swarm had selected this destination from the departure site, by way of the scout bees."

## Finding a New Home by Dancing

A century later, Martin Lindauer discovered the details of how the scouts communicated their recommendations to the swarm. In the spring of 1949, Lindauer observed a honey bee swarm that had settled in a bush, and to his surprise saw that many workers performed waggle dances (a communication system known to convey information about food source location; see chapter 5). Thinking at first that these bees might indicate some profitable flower patches, perhaps to supply the swarm with nutrition until it had found a suitable new site, he noticed something strange. Several of the dancing bees seemed covered in black dust, and it wasn't the black pollen of poppy flowers—when Lindauer caught them and held them close to his nose, he discovered that these bees smelled of soot. It dawned on Lindauer that these "dirty dancers" had actually returned from a disused chimney in war-destroyed Munich—that they had been scouting for nesting sites, not flowers. He had discovered that the same "language" that bees use for communicating flower locations is also used when a swarm of honey bees searches for a new home.

The history of this discovery is an interesting case study of how careful observation can lead to insights that would be all but impossible with any form of automated data recording, such as "ethomics," by which people hope to classify, using multi-camera systems, motion capture, and artificial intelligence, the entire behavioral repertoire of animals. It is very hard to conceive of an artificially intelligent agent that would (without being intricately programmed for it) not only have the inquisitiveness to be puzzled by black-dusted dancers, but have the intuition to sniff them, and to draw the correct insight about the target (and its intended function) that the bees had indicated. Human curiosity, paired with a scientist's careful observation and insight, is hard to beat. Building computers that are good at chess is a trivial task in comparison.

Lindauer found that many dozen swarm scouts explore a territory of up to 70 km$^2$, and return to the swarm communicating the coordinates of any suitable cavity they have discovered, using the dance "language" (figure 8.5). Dance followers decode the spatial information presented to them, memorize it, and then fly to inspect an indicated location themselves—a highly specialized form of social learning. There is plenty of conflicting information: different scouts return with information about different sites and their varying quality. Over several hours or indeed days, however, a consensus is built, and in the end all the dancers in the swarm appear to agree on the same location: it is at this point that the swarm lifts off and moves to the new agreed-upon home, which can be several kilometers away. When Lindauer first reported these observations to his mentor Karl von Frisch, the latter exclaimed: "Congratulations! You have witnessed an ideal parliamentary debate; your bees can evidently change their decision when other scouts announce a better nesting site!"

## Democratic Decision Making in the Honey Bee Swarm

A generation later, Tom Seeley continued the work on bee swarming behavior—indeed, he dedicated decades of his research career (and a whole book, memorably entitled *Honeybee Democracy*) to the process of how the decision on a new home is reached. He discovered that, remarkably, the organization of the consensus-building is entirely decentralized: no one individual counts the votes for the various indicated sites, nor is there a leader, nor

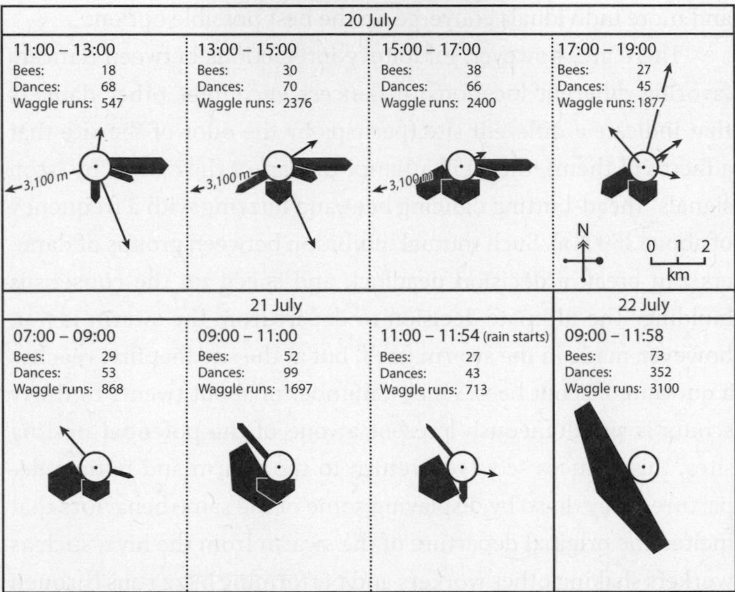


**Figure 8.5. Honey bee swarming.** ***Top:*** Small honey bee swarm clustered on a tree branch. Swarm sites such as this can persist for more than three days, during which time scouts investigate potential sites and communicate their findings by dancing on the swarm's vertical surface. ***Bottom:*** Decision making in a honey bee swarm, from the time the first potential nest site was advertised until eventual lift-off. Each panel shows a 2-hour interval, with observations over three days. Circle with central dot represents the location of the (stationary) swarm; arrows indicate the direction and distance of the potential nest site; width of arrows indicates the number of bees dancing for that nest site in the time period shown. Reprinted by permission of Seeley et al. (1999).

synchronous responses to an order; nor do individuals even compare the spatial or nest site quality information delivered in the dances. Instead, better locations are simply indicated by longer dances, thereby increasing the probability that a worker moving randomly on the swarm cluster is likely to bump into a dancer with better information by mere stochastic processes. More individuals will thus inspect better sites indicated by longer dances, and then return to the swarm cluster to indicate them in their own dances—ultimately resulting in a snowballing effect in which more and more individuals converge on the best possible option.

There are, however, inhibitory interactions between dancers favoring different locations: if dancers encounter other dancers that indicate a different site (perhaps by the odor of the site that adheres to them), they will attempt to disrupt the dances by "stop signals" (head-butting dancing bees and buzzing with a frequency of about 350 Hz). Such mutual inhibition between groups of dancers can break a decision deadlock and speed up the consensus building. The ultimate decision to depart from the swarm is not, however, made in the swarm itself, but at the site that first reaches a quorum of scout bees. Once a number of about twenty to thirty scouts is simultaneously present at one of the potential nesting sites, this induces scouts to return to the swarm and initiate departure. They do so by displaying some of the same behaviors that incited the original departure of the swarm from the hive, such as workers shaking other workers and performing buzz runs through groups of inactive individuals.

Once the swarm has been set in motion, knowledgeable scouts (that had earlier inspected the target site) act as streaker bees to guide the swarm in the right direction. They do this by performing conspicuous high-speed flight movements near the top and front of the swarm, then falling back by slower and less visible flight into the swarm, only to reinitiate another high-speed flight segment pointing in the right direction. Curiously, human crowds can be steered in a similar manner by only a minority of informed individuals. In such crowd movements, each individual needs only to adjust their movements to the nearest or most conspicuously moving individuals (no one needs to count the numbers of individuals moving in any direction)—and the result is that the intended movement soon spreads to all individuals of the "swarm."

## Mindless Individuals in the Intelligent Swarm?

In such collective decision-making processes, where individuals function as information gatherers or guides for the group, there is a synergy effect in that the swarm appears to "know more" than the sum of its individual members. The behavior of a bee swarm (or any functioning social insect colony) can be so tightly integrated, so well-coordinated, that at times it resembles a single being. This is captured by the term *superorganism*, in which the specialists of the social insect colony are likened to the cells and organs of one multicellular animal. In this view, the individual scouts of the swarm have functional similarities to the sensors and nerve cells that collect and process information about the world in an animal's brain, as Tom Seeley has highlighted in his book *Honeybee Democracy*. Social bees are special in this regard because of the high degree of relatedness of individuals of a colony: all workers are daughters of the same mother. This relatedness means that individuals gain unusually high fitness benefits from tailoring their information sharing so that it serves the entire colony, which in turn has facilitated the evolution of the many unique forms of communication we have encountered in these pages.

Based on such observations, a popular view holds that in social insects, the individual is a mindless machine, and intelligent behavior only emerges by self-organization as a function of the group. This has led to the notion, sometimes expressed in whispers, that swarm intelligence is somehow generated by a qualitatively different form of mind than the individual, a "collective mind." It is useful in this context to distinguish the real psychological concepts that deserve the terms "mind" and "intelligence" from their metaphorical use. Only individual living beings have minds, or intelligence. Individual social insects, like individual humans, have dedicated socio-cognitive processes that facilitate group coordination (though the processes and outcomes differ profoundly between us and them), which in turn can benefit the group as a whole, as well as its individual members. In both humans and bees, this allows solving tasks that no individual could solve, and indeed tasks that are unlikely to be solvable at all without coordination between individuals. In humans, as in bees, collective enterprises might *appear* to an outsider as a form of swarm intelligence that necessitates a collective mind.

But the memory for particular nesting sites, for instance, is stored in the brains of only those select individuals that have inspected the site, or learned of it from these knowledgeable individuals via the dance. Even if there should be a unique emotional state linked to swarming, as Buttel-Reepen suggested, this state is still one of individuals, not of the swarm as a collective being. There is nothing it feels like to *be* the swarm, and hence there is no collective mind. It only feels like something to be an individual *within* the swarm, and there are as many minds, and experiences, as there are individuals within the group. Individuals can certainly cooperate, and even where they compete—looking at humans now—the collective efforts of many individuals can result in such magnificent outcomes as the Manhattan skyline. But the intelligence, the minds and the cognition still reside within individuals' brains, and we know at this stage of the book that individual bees are certainly cognitively capable. The collective problem-solving strategies that we observe in social insects have evolved over many generations, not by swarms coming up with intelligent solutions to novel challenges.

We have learned in this chapter that bees have the crucial cognitive abilities both to "invent" new foraging techniques and to learn from each other, facilitating the cultural spread of such techniques. It has not yet been demonstrated that such abilities can prevail over multiple biological generations. It is possible that the interruption of the active foraging process in winter (as seen in temperate bumble bees and honey bees) might prohibit the transmission of cultural information across the years. Tropical social bees, in which flight activity persists year-round, might be the best candidates for investigating cultural processes.

Could there be cumulative culture in bees, where a behavioral innovation builds on an earlier one that is already widespread in a population? It is conceivable that, let's say, a culture of string-pulling bees might subsequently apply their skill toward a wholly new task—but it is not actually easy to conceive of real-world challenges for which this skill might be truly beneficial. Thus, the absence of a particular behavioral capacity in wild animals is not evidence that the ability is "hard to evolve," or that those animals lack adequate intelligence, but might in many cases simply reflect the absence of relevant natural challenges.

We have by now discovered many of the bees' outstanding individual and social learning capacities. But how can all these abilities be accommodated in the tiny microcomputers that are bees' brains? This is what the next chapter is about.

# 9 The Brains behind It All

> The excellence of the psychic machine does not increase with zoological hierarchy; instead one realizes that in fish and amphibians the nervous centers have undergone an unexpected simplification. Of course their grey matter has increased considerably in mass; but when the structure of their brains is compared with that of bees or dragonflies, they are excessively plain, coarse and rudimentary. It is as if one were to pretend to hold as equals the merits of a rough grandfather clock with the quality of a fine pocket watch, a marvel of fineness, delicacy and precision. As always, in building her marvelous works, nature distinguishes herself much more in her tiny creations than in the large.
>
> **—Santiago Ramón y Cajal and Domingo Sánchez y Sánchez, 1915**

I wish I could tell you a crisp story along the lines of, "The bee brain is small and simple, and therefore it's an easy stroll to explore how it works." The epigraph informs you that I'm unlikely to serve you this narrative. The bee brain is indeed compact, but because of the behavioral complexity that can be generated from within this tiny biocomputer, the bee has long been a model system for exploring how brains work. In fact, we will learn from the historical excursion in this chapter that some key breakthroughs in brain science were made first in bees before being extrapolated to humans and their mammalian relatives.

The first author of the quote above, Cajal (1852–1934), was a pioneer of the exploration of brains and was clearly impressed with the complexity of the insect brain in comparison with those of some vertebrates. He showed early promise as an experimenter when, at the tender age of eleven, he managed to manufacture gunpowder, and from scraps, a significant-sized cannon, with which he promptly annihilated the neighbor's freshly constructed garden gate. The detonation earned the preteen a three-day prison sentence, but he went on to win the Nobel Prize in Physiology or Medicine in 1906 and earned a reputation as a founding father of neuroscience.

One of the key ideas Cajal promoted is the "neuron doctrine"—the notion that the nervous system is not a body-wide, continuous network, but is made up of discrete units, the neurons (nerve cells), which communicate with each other across connection points called synapses. There are about 850,000 nerve cells in the bee brain. For comparison, there are about 100,000 times more in humans (86 billion). However, neuron numbers tell us nothing about intelligence or computational complexity. A bucketful of transistors is not more complex than a handful. What matters is how the individual elements of the circuits are connected. Before we explore how the bees' relatively few neurons might be wired up to support the impressive cognitive capacities that we have witnessed, let me introduce the insect brain's gross neuroanatomy.

## The Basic Construction Plan of the Bee Brain

Like ours, the brain of the bee is bilaterally symmetrical. The largest and main part of the insect brain is the *protocerebrum*. It contains the optic lobes, which process information from the compound eyes, and the mushroom bodies, prominent dorso-frontal structures with important functions in integrating information from multiple sense organs, as well as in learning and memory. The protocerebrum also features an unpaired structure, the central complex, which contains the computational centers for integration of the polarized light–based sky compass, information about the animal's own position and movement (as needed in path integration; see chapter 6), and landmark information. The central complex also controls motor commands sent to other nerve centers, which in turn control the movements of legs and wings (figure 9.1)—and it has even been suggested as the neural structure mediating conscious experience in insects.

The paired antennal lobes are the principal olfactory relays: they process chemosensory information from the antennae. The optic lobes contain three *ganglia* (distinct nerve clusters, typically containing both nerve cell bodies and connections between them, with a variety of computational functions): the *lamina*, *medulla*, and *lobula*. The visual system makes up about half of the entire brain. One might intuit that it makes sense for it to be so large, to support many of the functions in flower and visual pattern recognition we have just discussed. But Cajal's wise words about grandfather clocks and pocket watches should remind us that the size of any device is a hopelessly inadequate measure of its function, its complexity, or its inner workings. The truth, as we will soon realize, is that we have no idea why the bees' optic lobes need to be

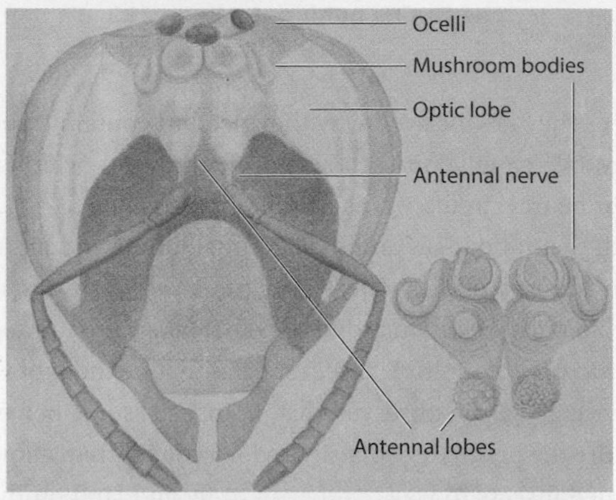


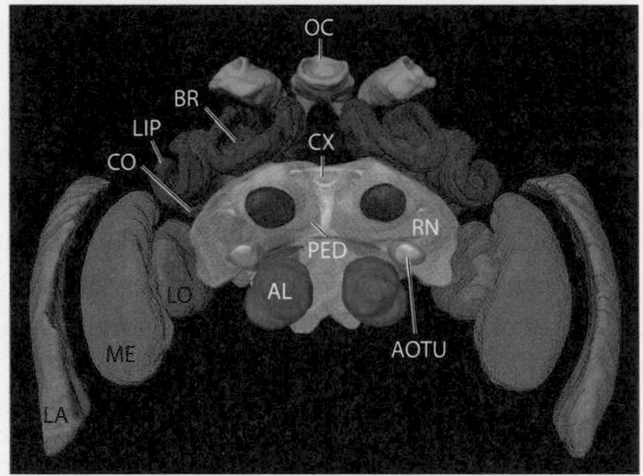


**Figure 9.1. Frontal views of bee brains, 1850 and 2021.** ***Top:*** The world's first drawings of the honey bee brain displayed in a transparent bee head, by Félix Dujardin. Shown are the convolutions (calyces) of the mushroom bodies (which Dujardin discovered), including the three dorsal ocelli and the antennal lobes (with olfactory nerves). The inset shows the regionalization of the mushroom bodies and the antennal lobe glomeruli. ***Bottom:*** Reconstruction of a bumble bee brain via micro-CT, a computerized X-ray imaging procedure. The optic lobes (shades of yellow) contain nerve centers called lamina (LA), medulla (ME), and lobula (LO). They send information to other brain areas via the anterior optic tubercle (AOTU) in two parallel pathways. The mushroom bodies (red) are centers of learning and memory, subdivided into the collar (CO), lip (LIP), basal ring (BR), and pedunculus (PED) regions. The antennal lobes (AL) are the brain's first olfactory processing relays. The ocellar (OC) neural networks (turquoise) are shown at the top. Remaining neuropils (RN) are rendered transparent to make the central complex (CX) visible.

as large as they are, or even why they need to contain three rather than, say, two ganglia on each side of the brain—nor do we know why their neural circuitry is as complex as it is.

Both the optic lobes and the antennal lobes send information to the mushroom bodies. These structures were first described by French biologist Félix Dujardin (1801–1860), who compared the brain structures of multiple insect species that he thought varied in intelligence. Dujardin observed that, in insects, some behaviors do not require the brain: their distributed nervous system allows some decapitated insects to walk, and even to fly and land on their feet. He contrasted this with those insects that construct nests, provision their young, and have "memories of places and things already seen." Remarkably, in the honey bee, he suggested that one sign of their intelligence is their communication about flower locations (one wonders where he got that idea, in 1850). All this, he mused (correctly), wouldn't be possible without the head (or brain).

Dujardin then dissected the brains of a variety of insects, looking for correlates of the abovementioned behavioral capacities. He discovered brain compartments that were widely different in relative size and structure between the examined species—he labeled them "*corps pédonculés*" and pointed out that they resembled certain mushrooms. *Corps pédonculés* didn't stick, but "mushroom bodies" did. He found that these structures had elaborate, regular convolutions in hymenopteran insects, and likened these to similar convolutions in the mammalian cortex (figure 9.1, top). Dujardin concluded that "As intelligence dominates over instinct, the *corps pédonculés* and the antennal lobes tend to become considerably larger relative to the overall brain volume, as one sees when comparing cockchafers to locusts, ichneumon wasps, carpenter bees, solitary bees, and finally social bees, where the *corps pédonculés* represent a fifth of the brain and 1/940th of the body; whereas in cockchafers, they represent less than 1/33,000th of the body."

Not only Dujardin's explorations of brain structure and his detailed drawings are extraordinary. He also preempted (by well over a hundred years) the now fashionable approach of comparing not absolute but relative brain sizes in proportion to body size (brain-to-body-mass ratio), as well as the relative size of certain brain areas compared to the entire brain. He also attempted to link such measurements to animal intelligence. Like many of today's scholars, he was a little vague on what exactly constitutes intelligence, or how one might measure it. The quotation above *appears* to indicate a progression of mushroom body volume that culminates in social bees (which are implied to have the highest intelligence), but Dujardin's published work contains no evidence that honey bee mushroom bodies are any larger than those of solitary bees. We now know that *all* nest-constructing, brood-provisioning Hymenoptera (including solitary ones) have enlarged mushroom bodies with elaborate convolutions, compared to their vagabond relatives (see chapter 5). It was the evolutionary transition from a homeless insect to one that needed to remember where home (and its brood) was, and the challenges of hunting for adequate nutrition within range of that home, that corresponded to an increase in mushroom body volume.

Counter to Dujardin's thinking, the evolution of sociality, or the honey bee's communication system, does not seem to have gone hand in hand with substantial changes in brain structure. Even unique evolutionary innovations clearly linked to sociality, such as the honey bee dance language, have no obvious correlate in gross neuroanatomy. For example, there is no specific "dance module" that would distinguish the brain of the worker honey bee from that of related species that lack the dance language. All the other neurobiological differences that no doubt exist between, for example, social and solitary bees (and dancing and non-dancing bees) must be sought inside the fine details of neural circuits, not in the size of the larger elements of the brain.

## The Discovery of Nerve Cells inside the Bee Brain

American writer and biologist Frederick Kenyon (1867–1941) was the first to explore the inner workings of the bee brain. His 1896 study, in which he managed to dye and characterize numerous types of nerve cells of the bee brain, was, in the words of the world's foremost insect neuroanatomist, Nick Strausfeld, "a supernova." Not only did Kenyon draw the branching patterns of various neuron types in painstaking detail, but he also highlighted, for the first time in any organism, that these fell into clearly identifiable classes, which tended to be found only in certain areas of the brain.

One such type he found in the mushroom bodies is the Kenyon cells, named in his honor. Their cell bodies—the part of the neuron that contains the chromosomes and the DNA-decoding machinery—are in a peripheral area enclosed by the *calyx* of each mushroom body (the mushroom's "head"), with a few additional ones on the sides of or underneath the calyces (figure 9.2). A finely arbored dendritic tree (the branched structure that is a nerve cell's signal "receiver") extends into the mushroom body calyx, and a single axon (the neuron's "information-sending output cable") extends from each cell into the mushroom body *pedunculus* (the mushroom's "stalk").

Extrapolating from just a few of these characteristically shaped neurons that he could see, Kenyon suggested (correctly) that there must be tens of thousands of such similarly shaped cells, with parallel outputs into each mushroom body pedunculus. (In fact, there are about 170,000 Kenyon cells in each mushroom body.) He found neurons that connect the antennal lobes (the primary relays processing olfactory sensory input) with the mushroom body input

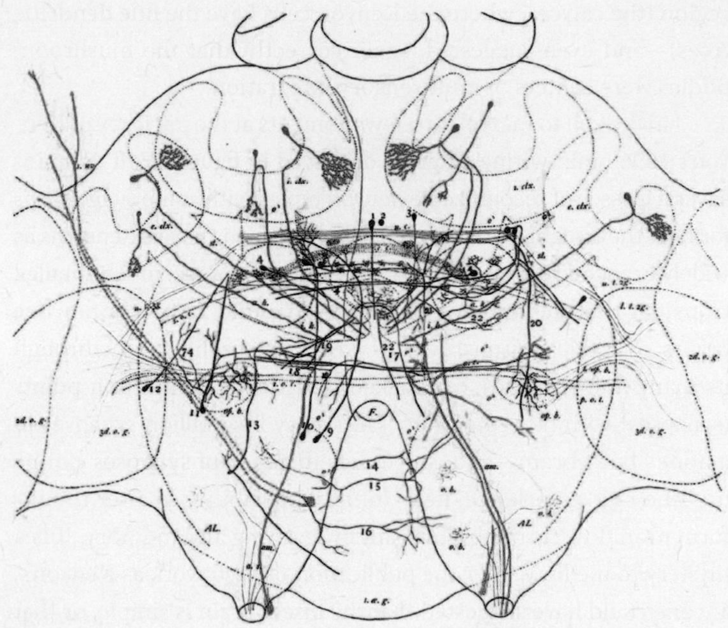


**Figure 9.2. Frontal view of the bee brain with several distinct types of neurons** (from F. Kenyon, 1896). Kenyon cells (in red, inside the mushroom bodies): a clearly recognizable type of cell with its dendritic tree in the mushroom body calyx (labeled "clx"). On the far left and right are the lobula (3rd optic ganglion = "3d. o. g.") and medulla (2nd optic ganglion = "2d. o. g."). At the bottom are the cut antennal nerves, and the antennal lobes ("AL," shown right above the cut antennal nerves). The central complex (including the protocerebral bridge, fan-shaped body, and ellipsoid body) is also shown at the center of the brain (see figure 9.5 for further detail). Several neurons bridging the two hemispheres are shown in black.

region (the calyces, where the Kenyon cells have the fine dendritic trees)—and even suggested, again correctly, that the mushroom bodies were centers of multisensory integration.

I invite you to marvel for a few moments at the intricacy of Kenyon's 1896 brain wiring diagram displayed in figure 9.2. It contains several classes of recognizable neuron types, with some suggestions for how they might be connected. Many neurons have extensions as widely branched as full-grown trees—only, of course, much smaller. Consider that the drawing only shows around 20 of a honey bee brain's ~850,000 neurons. We now know that each neuron, through its many fine branches, can make up to 10,000 connection points (synapses) with other neurons. There may be a billion synapses in a honey bee's brain—and, since the efficiency of synapses can be modified by experience, near-infinite possibility to alter the information flow through the brain by learning and memory. It is a mystery to me how, after the publication of such work as Kenyon's, anyone could have suggested that the insect brain is simple, or that the study of brain size could in any way be informative about the complexities of information processing inside a brain.

Kenyon apparently suffered some of the anxieties all too familiar to many early-career researchers today. Despite his scientific accomplishments, he had trouble finding permanent employment, and moved between institutions several times, facing continuous financial hardship. Eventually, he appears to have snapped, and in 1899 Kenyon was arrested for "erratic and threatening behavior" toward colleagues, who subsequently accused him of insanity. Later that year, he was permanently confined to a lunatic asylum, apparently without any opportunity ever to rehabilitate himself, and he died there more than four decades later—as Nick Strausfeld writes, "unloved, forgotten, and alone."

It was not to be the last tragedy in the quest to understand the bee brain.

## Neurons Processing Visual Information in the Bee Brain

It is unlikely that Kenyon ever knew of the impact he had on Cajal's work on the other side of the Atlantic. Cajal was so excited by Kenyon's discoveries that they inspired him to explore the insect nervous system in more detail. Jointly with his colleague Domingo Sánchez, he focused on the visual system (the three ganglia: lamina, medulla, and lobula) of several insect species, including the honey bee.

Recall that the insect compound eye is made of thousands of individual eyelets, each with a hexagonal lens as the external face of a structure called an ommatidium underneath. This structure contains the photoreceptors, which can be sensitive to light of different wavelengths. Different species of bees have between 1,000 and about 16,000 ommatidia (honey bee workers have ~5,500; see chapter 1). UV and blue photoreceptors send axons (neuron cables) all the way to the medulla ("long visual fibers"), whereas green receptors terminate in the lamina ("short visual fibers"). The optic ganglia are each organized into "columns" (or "cartridges")—in fact, in the lamina and medulla there are as many cartridges as there are ommatidia. In the lobula, the number of cartridges is somewhat lower. Both the medulla and the lobula send axons to the central brain. The columns in the optic ganglia are highly repetitive in that they contain replicates of the same neuron types in each column.

The diversity of neuron types stunned Cajal. He observed: "The complexity of the insect retina is something stupendous, disconcerting, and without precedent in other animals. When one meditates, finally, on the infinite number and exquisite adjustment of all these histological factors, so delicate that the highest powers of the microscope hardly bring them under observation, one is completely overwhelmed." We now know that even the humble fruit fly has over 150 neuron *types* just in its optical ganglia (we don't yet have the numbers for bees, but they're likely to be as high; for comparison, the human retina has fewer than 100). A single column of the medulla can have several dozen neuron types—and these types will be found (with some variation in fine branching structure) in every single column (figure 9.3).

Many of these neurons show connections perpendicular to the flow of information from the retina to the central brain, some for local comparisons between signals from neighboring ommatidia, for example for contrast detection and enhancement: after all, contrasts typically define the boundaries and thus the identity of things and living beings. Other neurons have wide "receptive fields," integrating information from entire eye regions, for example to measure the average brightness of a scene. These tangential neurons give the medulla a layered appearance—in bees, the medulla has eight such layers, whereas the lobula has six stripes, each made of multiple neural cross-connections (the human retina has a mere two layers of such lateral connections).

The medulla and lobula contain various neuron types described as "simple feature detectors," each of which analyzes a particular aspect of a scene or an object: color-coding or brightness-coding neurons, movement sensors, etc. The bees' lobula, for example, contains two classes of so-called edge orientation detector neurons. When measuring visual neurons' properties, researchers will typically place microelectrodes in or on neurons of interest and bombard the animals with a battery of visual stimuli—colored lights, dots moving in various directions, briefly flashed stimuli versus persistent ones, bars of various orientations and movement speeds—to pinpoint those neurons' response characteristics. The edge orientation detectors in the lobula respond most strongly to oriented

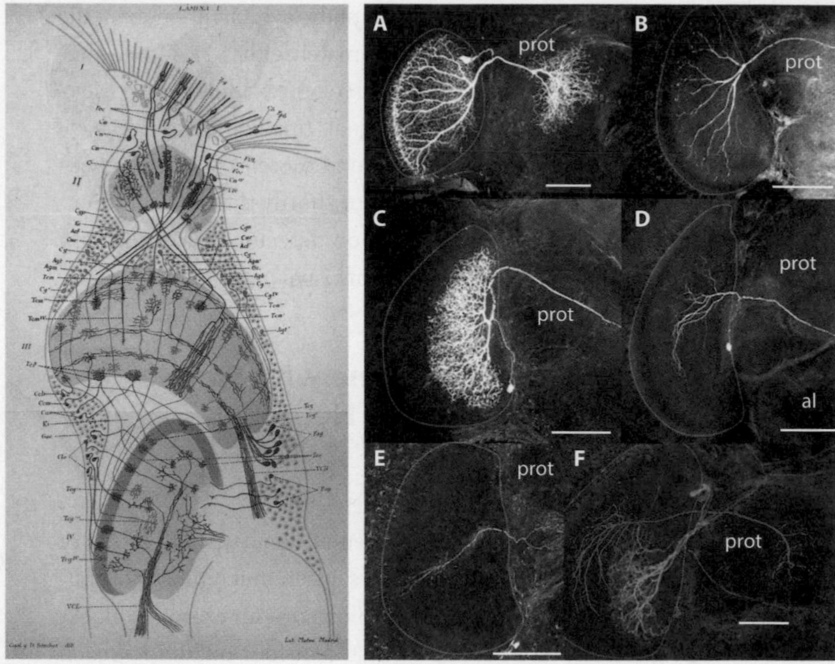


**Figure 9.3. Diversity of neuron types in the visual system of bees.** ***Left:*** Neuron types in the bee visual system (from Cajal & Sánchez's 1915 article). The fan-shaped area at the top is the eye's retina. The three shaded areas beneath are (top to bottom) the optic nerve centers lamina, medulla, and lobula, with two chiasmas (crossing-over points of nerves) between these ganglia. Recognizable neuron types are shown in different colors. Locations of cell bodies are indicated outside the neural relay areas where the nerve cells' input and output regions are connected. ***Right:*** Neuron types in one of the optic ganglia—the lobula—of the bumble bee, made visible with fluorescent dye, showing their connections with the protocerebrum ("prot")—central brain. Some neurons have extremely wide branching patterns (e.g., A, C) and receive signals from the entire eyes, whereas others are "columnar" (e.g., E), perhaps conveying information only about a few neighboring ommatidia (corresponding to "pixels") from the eye to the brain. Scale bars: 100 μm.

bars moving anywhere across the width of the eye, but are maximally sensitive to orientations angled approximately either +110° or −110° from the vertical. They will also respond to other, similar edge orientations, but less strongly. Most visual signals (including flowers) have edges, to which such neurons will respond more or less strongly as the bee scans over them. We will find out shortly that this combination of just two types of edge orientation detector neurons is very powerful at recognizing a large variety of visual patterns.

## How Much Can Be Done with Simple Feature Detector Neurons?

Over the years, researchers have presented bees with all kinds of challenging pattern discrimination tasks, often much more complicated than those typically presented by flowers—for example, black-and-white circular patterns that consisted of four quadrants, each quadrant having a different orientation of multiple stripes. Bees could manage these tasks just fine—but does this mean that they really remember the images in their full complexity?

Mark Roper on my team, working with Chrisantha Fernando from Google DeepMind, constructed artificial neural networks, modeled on bees' brains, which deployed only two simple feature detectors—that is, two kinds of lobula neurons, each of which is especially sensitive to lines or edges that run in a particular direction. These algorithms were capable of recognizing complex visual patterns, like a circle carved into four, with stripes running at different angles in each quadrant. So a bee could store these complex visual patterns just by memorizing the signals from these neurons—without actually storing "virtual images" in its memory, in other words, without an awareness of the actual patterns. In fact, the models show that by using such feature detectors *only*, a bee might actually perform better than has been found in empirical tests. This is despite the fact that these models are so simple as to be caricatures of the real complexity of visual processing. They employ just two of the several dozen types of neurons of the lobula, and the number of synapses (contact points between neurons) in the models is diminutively small compared to the intricate connectivities in real neural circuits.

The diversity and complexity of neuron types in the visual system of the bee contrast sharply with the extreme simplicity of circuits minimally required for many seemingly clever cognitive tasks. For example, a simple four-neuron network can support the counting abilities of bees (see chapter 6), provided the bees inspect the to-be-counted items one by one, sequentially. If you ask computer scientists for the simplest neural network with which a given cognitive ability can still be implemented—say, learning sequences of landmarks, path integration, or foreseeing the outcomes of one's own actions (a consciousness-like phenomenon)—the answer is typically that this can be done with dozens to, maximally, some hundreds of neurons. Thus, a big mystery about bee brains and their cognition is not "How can animals with such small brains as bees do so many clever things?"—but the opposite: "Why does any animal need as large a brain as a bee's?"

## A Single Neuron that Can Learn

For the bee to learn about flower features (such as colors and scents), there needs to be a neural pathway that responds to the sugar reward contained in floral nectar, and that sends its signals to the same regions of the brain where information from the visual and olfactory sensory periphery is processed. Martin Hammer, a student of Randolf Menzel, described such a pathway in the honey bee brain—in the form of a single nerve cell, which he called VUMmx1 (for ventral unpaired median maxillar 1; figure 9.4). This neuron is perhaps one of the most widely branched ones in the bee brain; its cell body is located in the subesophageal ganglion (close to the mouthparts, where it receives input from the bee's sugar receptors). It sends its branches (and thus, its information) into the antennal lobes and also into the calyces of the mushroom bodies, which additionally receive olfactory information (from the antennal lobes) and visual information (from the medulla and lobula).

Martin Hammer discovered that stimulation of this single neuron electrically would fool the bee into thinking that it had just received a sugar reward. This means that such bees could learn about floral odors without actually having received a reward: if the bee received an odor at the same time as the VUMmx1 neuron was triggered by the experimentalist, the bee would learn the odor as if it had actually received a sugary reward. While this is not definite proof that this neuron is the *only* one representing the reward pathway in bees, this appears at least plausible: the neuron's experimental electrical stimulation has the same effect as the bee finding sweetness with its proboscis. Consider, in comparison, that the dopamine reward system of mammals contains tens of thousands of nerve cells all conveying the same message. The insect brain has no room for such profligacy, and in some cases a function may indeed be mediated by a single cell.

But isn't this risky? What happens if something untoward should befall that single cell? If one function is underpinned by just a single mechanism, without any backup, then its damage would undoubtedly spell disaster—but it might well be that the worker bee is too short-lived for evolution to provide for such eventualities.

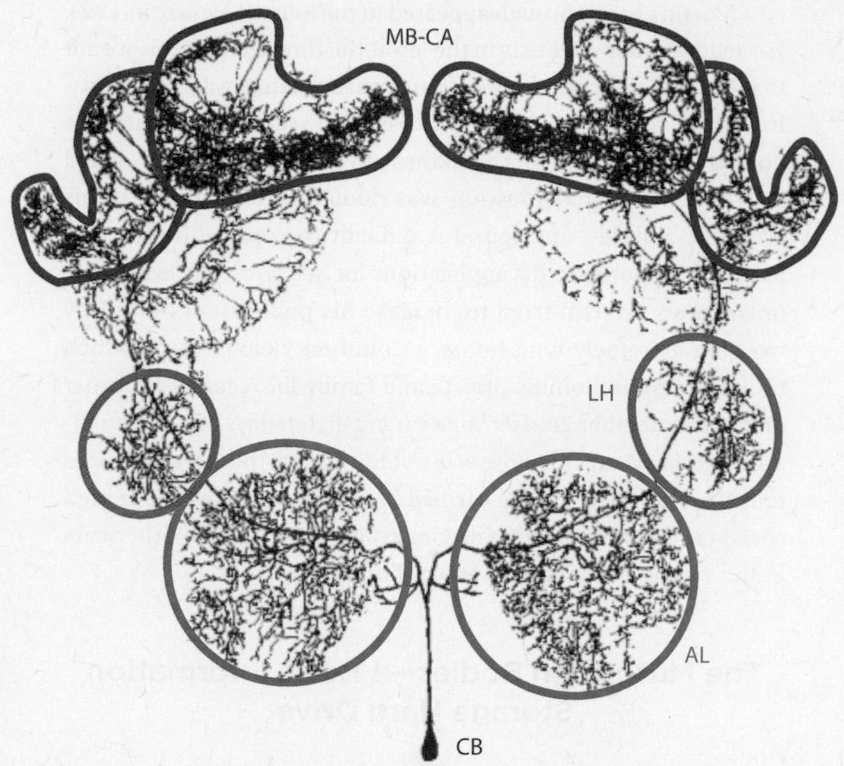


**Figure 9.4. The intricate structure of a "reward neuron" in the bee brain.** This so-called VUMmx1 neuron sends a "sweetness signal" to many regions of the brain. Its cell body (CB) sits in the subesophageal ganglion of the brain; its branches extend throughout the antennal lobes (AL), the lateral horns (LH), and then to the mushroom body calyces (MB-CA). Artificial stimulation of this neuron can cause the bee to "think" it has tasted sugar, and the bee can learn to link this illusory sensation to an odor presented at the same time. Reprinted by permission from Hammer (1993).

Martin's breakthrough appeared in the journal *Nature* in 1993. His fellow junior scientists in the lab at the time felt that this meant that he had "made it"—that his career in academia was a certainty. It was not to be. He was not successful in securing a permanent post in the years after his breakthrough publication. He struggled intermittently with depression, was riddled with doubt about his scientific abilities, and found it difficult to cope with rejection letters in response to his applications for academic posts at other universities. Martin tried to increase his publication output by working extremely long hours, a common vicious cycle, which took a severe toll on his private and family life. Martin Hammer died on September 24, 1997 in a car crash, ten days after his fortieth birthday. A suicide note was not found, but the circumstances told their own story—the car had been driven into a tree at high speed under unchallenging driving conditions, with no other cars in the vicinity. His seatbelt was unbuckled.

## The Mushroom Bodies—a Bee's Information Storage Hard Drive

Neuronal wires coming in from the visual and olfactory sensory periphery connect to the mushroom bodies' Kenyon cells—the connection points are visible as microglomeruli (complexes of synapses) under the microscope (figure 10.7 in the next chapter). Importantly, the VUMmx1 reward neuron discovered by Martin Hammer also connects to the same input region of the mushroom bodies. Thus the sensory pathways signaling "sugar" as well as those conveying flower colors and scents all converge here. Even more importantly, the microglomerular connections are "plastic"—they change when the animal learns, so that when there are simultaneous signals from the reward pathway and a stimulus such as a color signal from a flower, these become represented in the neural network as new or strengthened synaptic connections. The mushroom bodies are thus a neural information storage device.

The immense storage capacity of this "neural hard drive" comes about because the mushroom body uses a principle that is also used in machine learning: a so-called "fan-out, fan-in architecture." The numbers of neuronal "wires" from the antennal lobe (olfactory pathway) and the visual system are only in the hundreds—but they connect to 170,000 Kenyon cells, and these are then "read out" by approximately 400 mushroom body extrinsic neurons, which connect to regions of the central brain from where behavioral responses are selected and coordinated. The connections between sensory projection neurons (which convey signals from sense organs to the mushroom bodies) and Kenyon cells are relatively sparse: each Kenyon cell is thought to be innervated by only about ten sensory projection neurons. The result is a so-called "sparse code"—where only a tiny fraction of the Kenyon cells is activated for each single incoming sensory stimulus. This sparse code results in high specificity: even two similar visual scenes seen by the bee can result in completely different activation patterns of the Kenyon cells. This type of coding results in extremely high memory capacity.

This is illustrated by a model of landmark orientation in ants, where the authors modeled the described "fan-out" architecture from 360 visual projection neurons to 20,000 Kenyon cells. Even though the number of Kenyon cells was much smaller than in bees, the model could store 350 realistic visual scenes from the ant's cluttered natural environment without the ants confusing any of the scenes. When the number of memorized visual scenes was just 80, the computer-modeled "ants" could still recognize each of them (with minimal error rates—about seven times lower than chance) even when displaced 25 cm from the location from which they memorized the scene. Again, it has to be borne in mind that such models involve extreme simplifications of the real neural circuitry in place, so the actual memory capacity of bees (and ants) is likely much higher.

So, if anyone ever tells you that the small brain of insects constrains them to have little memory capacity, or that a bee can keep "only one thing in mind," you can now set them right.

## Complex Learning with Simple Brain Circuits

My PhD student Fei Peng, a classical psychologist by training, honed his brain-modeling skills while working on this ant study. He subsequently built a model of the bee's mushroom body to explore how the complex flower scent–learning abilities of these insects might come about. Specifically, bees can respond in highly differentiated ways when they are faced with multiple, potentially conflicting scents (where some may be linked to reward and others to no reward or to distasteful food). One such psychological phenomenon is *peak shift*, in which animals not only respond to a previously rewarding scent, but respond even more strongly to odors that are similar to the rewarding stimulus but even further from (more unlike) non-rewarding ones. This has been regarded as a form of rule learning, in that bees don't simply respond to single previously encountered stimuli, but instead combine information about different scents to infer what is the best possible scent. Bees also display negative and positive "patterning discrimination," readily learning to respond in one way to a mixture of two odors and in the opposite way to each of the component odors on its own. For example, in positive patterning discrimination, they have to learn that a combination of rose scent and geranium scent is rewarding, but, presented individually, neither rose scent nor geranium scent is rewarding. Such phenomena have historically been regarded as higher forms of intelligence than simple associative learning.

Fei built his mushroom body model based only on known details of neural olfactory information processing in the bee brain—for example, the established fact that synapses at the interface between neurons projecting from the bee's antennal lobes and the mushroom body's Kenyon cells can be altered by learning associations between odors and sugar rewards. But—even though none of the aspects of the model were "tweaked" to generate the kinds of complex learning phenomena described above, they "popped out" of the model. The mushroom body model performed much more complex operations than were put into it.

Fei discovered that the neuron circuitry mediating "simple" associative learning can also replicate the various more "intelligent" forms of learning, and can effectively multitask—replicating a range of different learning feats—without these abilities actually being built into the model. These findings not only question the notion that forms of learning regarded as "higher cognition" are computationally more complex than "simple" associative learning. Even more important, perhaps, these results show that a neural structure evolved to solve just the basic challenge of learning associations might spontaneously generate more-intelligent forms of learning without further evolutionary fine-tuning.

## The Insect Central Complex—a Sophisticated Navigation Device

In addition to the mushroom bodies, another important structure of the insect brain in which multiple sensory pathways converge, and which also stores memories, is the so-called central complex.

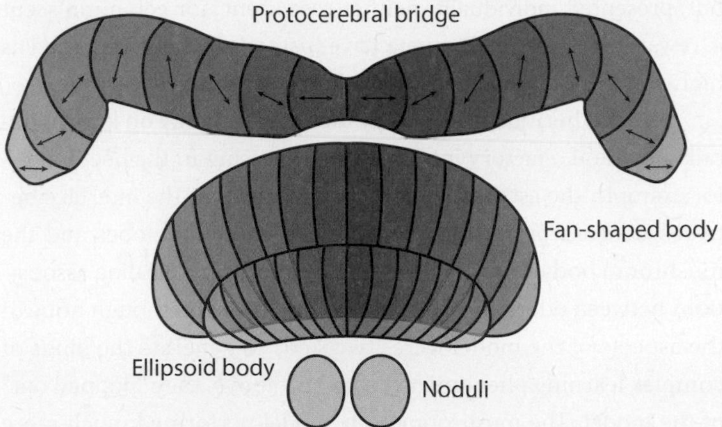


**Figure 9.5. The insect central body and its main compartments:** the protocerebral bridge (PB), fan-shaped body, ellipsoid body, and paired noduli. Nerve cells in the PB are tuned to different orientations of polarized light (as indicated by double-headed arrows), so that the PB as a whole contains a map of polarization directions. The PB, fan-shaped body, and ellipsoid body are organized in columns. Neurons from particular columns in the PB tend to be wired up with specific columns in the fan-shaped body or ellipsoid body.

This beautifully regular structure is common to all insects and is similarly structured in insects with highly divergent lifestyles, indicating that its functions across insects are at least somewhat similar and ancestral.

The central complex is the only unpaired structure in the insect brain, and as the name suggests, has a central position both geometrically in the brain and functionally in terms of sensory integration and behavioral action selection. It has four main compartments—the *fan-shaped body*, the *ellipsoid body* (together called the "central body"; see figures 9.1 and 9.5), the *protocerebral bridge,* and the *paired noduli.* Like many brain structures across animals (and also like the optical ganglia in insects), the central body is organized in columns—local neural networks that are highly repetitive in their internal wiring. Sixteen columns run through the protocerebral bridge and central body, and there are many layers of cross-connections between the columns as well. The central complex is multifunctional: studies from across insects indicate a principal function in navigation, including sun compass usage, path integration, and landmark memory.

Some components of the central complex deserve special attention because they are beautiful examples of the tight interaction between neural circuit structure and function. Neurons in the protocerebral bridge encode the direction of the sun and thus form a neurobiological compass, which is time-compensated (essential since the sun is only useful as a compass if one knows the time of day). What if the sun is not visible? We have learned in chapter 3 that bees are sensitive to polarized light. Since the pattern of polarized light in the sky moves in a predictable manner with the sun, this means that compass cues are available even when the sun itself is obscured by clouds. Neurons within each column of the protocerebral bridge are sensitive to one predominant plane of polarized light—but neurons in neighboring columns each respond to a different predominant plane (figure 9.5). As a whole, the protocerebral bridge thus contains a sophisticated compass device.

Beneath the protocerebral bridge is the fan-shaped body, one of whose key functions is in visual pattern memory. In fruit flies, at least, defined layers of neurons appear to code for, and memorize, features of a landmark, independently of the eye region on which it is viewed. Together, the protocerebral bridge and the fan-shaped body may thus allow an insect to link landmarks to compass cues, allowing a bee to select, for example, the correct direction home when it sees a familiar panorama.

The ellipsoid body likewise has a key function in orientation: a neural activity peak ("bump") rotates around the perimeter of the ellipse as an insect orients its body in different directions. The bump even indicates the correct previous direction when the insect walks in the dark, showing that the ellipsoid body uses input from the animal's own self-generated movements. The ellipsoid body allows the animal to maintain a direction even when external sensory input is absent, as well as keeping an update of the compass bearing when it decides to turn in a new direction in the dark. Finally, branches of neurons that measure travel distance have been found in the paired noduli of the central complex (figure 9.5).

In sum, the central complex collects information about movement distance, direction (from both external sensory signals and self-generated motion), and landmarks. It thus contains versatile navigational tools that allow such complex functions as path integration (which enables a bee that has searched for a novel floral resource along a tortuous path to return straight home, integrating data on all the distance segments traveled and angles turned along the outbound path—chapter 6).

## The Central Complex as the Seat of Consciousness?

Because the central complex integrates information from external stimuli, internal states, and past experiences, it could represent a kind of neural model of the familiar space around the insect, as well as of the self. In this view, bee scientists have even discussed whether it supports a form of consciousness in the insects. We will explore the psychological side of this topic in more depth in chapter 11. On the neurobiological side, support for the notion that the central complex mediates consciousness-like functions comes from parasitoid wasps that appear to deprive their prey of all self-motivated, goal-oriented behavior.

Rather than using the more widespread method of paralyzing its prey by injections into the nerve centers of the thorax (which control leg and wing movement), the jewel wasp, *Ampulex compressa*, stings its cockroach victims in the brain, and specifically in an area near the central complex. The prey is neither paralyzed nor deprived of its senses; instead, the injection turns it into a kind of zombie without any self-initiated actions. It can still walk—indeed, the wasp leads the docile cockroach to its burrow, where it will die a slow death by being consumed alive by the wasp larvae. The psychological and neurobiological effect of the venom of such "mind-controlling" parasitoids could potentially reveal a lot about the mind of the insect itself.

While one might assume consciousness requires a really large brain with a neocortex, this is not the case. First, one can never deduce the existence of any cognitive capacity from gross neuroanatomy: chimpanzees have Broca's and Wernicke's areas, brain regions that support language in humans but clearly not in chimps, so the presence (or indeed the absence) of a certain area tells us nothing about the existence of a cognitive capacity. Wholly different neural circuits can support similar behavioral abilities in different animals. Basic consciousness-like phenomena (such as predicting the outcome of one's own actions) can be implemented with just a few thousand neurons—not a prohibitively large number for an insect brain.

## Brain Waves in Bees

In terms of the question of whether insects "think," it is of particular interest to explore whether there is spontaneous activity in the brain. Any activity generated from "within the brain"—that is, in the absence of or distinct from external stimulation—

is potentially significant in the context of consciousness. Such activity that precedes any external stimulus occurs when animals focus their attention on certain aspects of their environment, anticipating something that *might* occur before it actually occurs. Attention allows animals to focus on important stimuli (such as a familiar flower, if you're a bee) and disregard others (such as unfamiliar flowers)—but it is more than simply a selective filter. The bee must know, somewhere deep inside the brain, what it's looking for, and feed this information to filtering mechanisms closer to its sense organs. The Australian neuroscientist Bruno van Swinderen and his team tested this by placing bees in a virtual reality environment that they could manipulate, and measuring their brain activity while they did so. His team found neural activity patterns that corresponded to paying attention to one or another object, and also found certain brain states that preceded the bees' selection of one or another stimulus.

Significantly, van Swinderen's team also discovered that insects have several types of neural oscillations ("brain waves"), including when they are asleep. Why does this matter for the question of consciousness? In humans, certain frequencies of brainwaves are linked to conscious states (as opposed to deep sleep or anesthesia). Conscious experiences need the integration of information from several different brain structures (including sensory areas, memory centers, structures that underpin motivation and action selection). It is thought that certain types of neural oscillations function to synchronize these brain areas' neural activity, so that information from them can be woven together into a coherent percept. The discovery of such oscillations in insects is thus exciting, especially in view of the fact that there are also distinct phases of oscillatory waves during sleep. The insect brain is thus never "switched off."

Bees have three distinct sleep phases. Deepest sleep is characterized by a distinctive crouching posture in which head, thorax, and abdomen are relaxed, antennae are immobile, muscle tonus and body temperature are decreased, and response thresholds to external stimuli are increased. Remarkably, exposing bees to odors during this sleep phase functions to consolidate experiences from the previous day in honey bees' memory. Could bees, like humans, have sleep phases in which memories are relived in dreamlike states? This might serve a function not only in memory consolidation, but also, since dreams stochastically shuffle bits of memory and thoughts into new configurations, perhaps in the exploration of possible real-world scenarios or alternative solutions to familiar problems.

## Different Lifestyles, Similar Brains

One curious observation is that related insect species with very different lifestyles often have very similar brains. Many impressive and unique behavioral or cognitive capacities, while they clearly must have a neural substrate, are not readily detectable in terms of gross neuroanatomy. For example, one of the most impressive recent discoveries in insect social cognition was Elizabeth Tibbetts and colleagues' discovery of individual face recognition in *Polistes* wasps. To explain briefly, some species of *Polistes* wasps have very small colonies in which each individual has distinct facial markings. Colonies are founded by several females which work out a hierarchy through extended duels; at the end, the winner monopolizes reproduction.

Wasps of a colony recognize one another, and know their place in the colony's hierarchy after determining their rank via fights with competitors. Because such fights are costly—they can cause injury or even death—they are best not repeated; it is useful to know one's place in the pecking order. Wasps may even learn about the fighting strengths of others by observation, and use "transitive inference": if one observes that individual A is stronger than B, and B stronger than C, then it follows that A is also stronger than C. Despite these abilities (quite possibly unparalleled in the world of insects), few discernible differences were found in the visual systems of these wasps from those of related species in which face recognition does not occur. This puzzle is mirrored in primates: the human brain, for all its obvious differences in cognitive output compared to other primates', appears, at least in terms of coarse organization, in many respects to be a scaled-up version of the primate brain.

This emphasizes that even seemingly major evolutionary innovations relevant to behavior and cognition may be generated by relatively small adjustments in neural circuitry that might be hard to detect, but relatively easy to evolve. Since many cognitive operations can be performed within fairly small neural circuits, if there is any selection pressure to evolve certain forms of intelligence, then those are likely to be found in the species in question. The absence of a particular behavioral capacity in wild animals is not evidence that the ability is "hard to evolve," or that a species lacks adequate levels of intelligence, but might in many cases simply reflect the absence of relevant natural challenges. For example, the reason that social bees do not recognize each other individually is not that it is not technically feasible with a small brain; rather, their individuals are too similar and too numerous for face recognition to be useful. Small changes in neural circuitry can generate large shifts in behavioral capacity, in part because existing circuits can often be co-opted with only minor modifications.

The study of bee brains has taught us that brains, even very small ones, are wired for cognition, for exploring the environment and extracting rules from it, for predicting the future, and for efficient information storage and retrieval. In this and the preceding chapters, we have learned about the various sensory and learning abilities of bees, and their bases in the nervous system. However, we have treated bees somewhat as interchangeable members of their species, without an exploration of their individual psychology. Yet psychological traits are by their very nature based on individual experiences and individually inherited modes of behavior, all of which differ from those of other individuals in a population. And so, in the coming chapter, we will explore whether bees can be said to have individual "personalities."

# 10

# "Personality" Differences between Bees

> We have now seen that insects do possess a decided preference for a number of successive visits to the same species of flower, although this is not invariably the case. . . . In [certain] cases humble-bees paid many visits to two different species of flower at the same time, passing alternately, without respect to colour, from one to the other after several visits. . . . These bees were a little more highly intellectual than their fellows, and could manage to work the two species together, although I should fancy more than two would puzzle them.
>
> **—Robert Christy, 1884**

It is obvious to any pet owner that different animals have different individualities—psychological traits that are recognizable in certain animals and distinguish them from others of the same species. Such traits can be the result of individual experiences, or genetic predisposition (inherited from parents), or a combination of both. But for many, insects of a species seem like indistinguishable, interchangeable, mass-produced entities—surely they can't be said to have "personalities"?

The first scientist to study individual differences in the psychology of invertebrates systematically was Charles Turner, who, as early as in his first paper on spider web construction (at age twenty-five, in 1891), observed pronounced differences in how individual web spinners coped with unusual geometric challenges, and referred to one individual spider as the "master mind of the locality." The identification of such individual differences, which he discovered in invertebrates as diverse as spiders, ants, and cockroaches, is a constant theme in Turner's work.

We have learned in recent years that in bees, differences occur in any psychological trait examined, and occur between individual bees (each of which will often respond similarly when tested repeatedly), as well as between colonies of bees in social species—unsurprisingly, since colonies are families of genetically related individuals. Different individuals have subtly different sensory equipment, which means they selectively perceive different aspects of their environment, and differences in brain structure, which determine that information is stored and used differently. Variation in individual intelligence is important for how well bees fare in the economy of nature, and variation between individuals of a colony determines the efficiency of their division of labor.

Variation between individuals and colonies can be heritable—for example, a colony of especially fast learners might pass this trait on to the next generation. Where psychological traits are heritable, they can be the raw material for evolution. If there is no heritable variation, there is nothing for selection to act on. For example, evolution can't easily make seven-legged insects, even if there should be an advantage to having an additional leg: there aren't typically seven-legged mutants around that might over time gain an upper hand over their six-legged cousins. On the other hand, as we will soon see, there is certainly heritable variation in psychological capacities such as learning facility in bees, and this means that learning-related traits can evolve rapidly over relatively few generations.

We will also discover in this chapter that not all variation between individuals is heritable. The dramatic differences between honey bee queens and their sterile workers concern every aspect of their sensory system, brain structure, and behavior—but this is not caused by any differences in their DNA, since these castes are not distinguished by their genes. Instead, the differences between queens and workers are epigenetic and are prompted solely by environmental factors (curiously, by the food they are given as larvae). We begin the chapter by introducing some of the technology that allows the quantification of individual specializations in behavior in social bee colonies.

## Using Microchips to Explore Bee "Personality"

The moment bees are marked in ways that make them recognizable as individuals (for example, with number tags; see figure 10.1), a wholly new perspective on their nature opens up. It becomes instantly obvious that different individuals of the same species behave very differently. Some bees are more aggressive than others, some are more hard-working, some more intelligent; some make fast and sloppy decisions while others are more careful, and so on. In recent years, quantifying such interindividual variation has been facilitated by new technologies such as RFID (radio frequency identification)—the same technology that is used in pet microchipping or season tickets in many public transport systems.

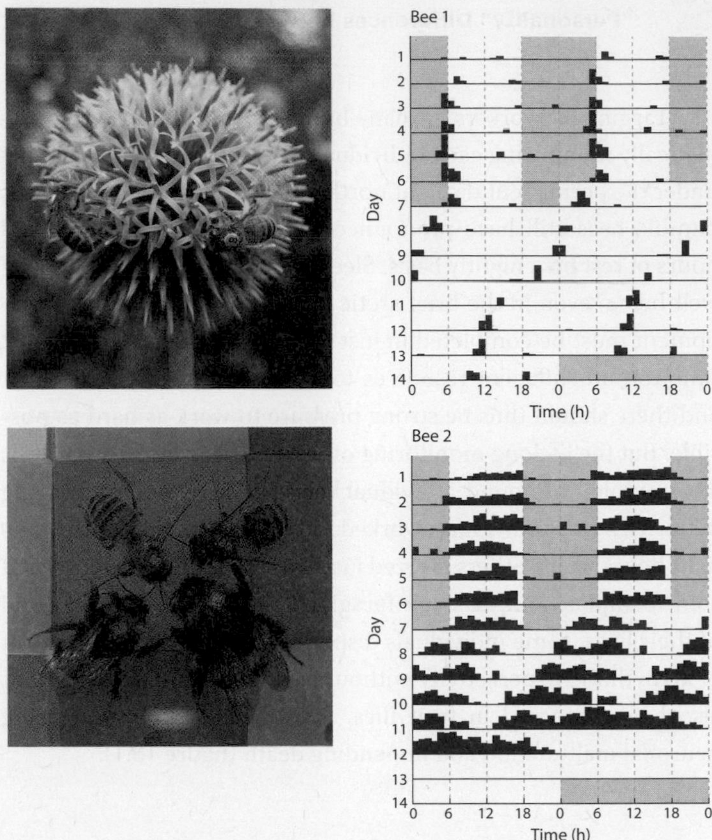


**Figure 10.1. Number tags and microchips reveal pronounced differences between individual bees.** ***Top left:*** When honey bee workers are fitted with number tags, they can often be seen to return to the same flower patches for several days, albeit with individually different temporal patterns, and visiting flowers in different sequences. ***Bottom left:*** Bumble bees and honey bees are marked with RFID (radio frequency identification) tags, which allows automated recording of their activity patterns. ***Right:*** Double-plotted activity diagrams ("actograms") of two individual bumble bees (height of bars indicates activity). Bees were first tested under 12 h light / 12 h dark conditions (as indicated by light-grey shading) and then under permanent daylight conditions such as found naturally under the midnight sun. ***Top right:*** Actogram of a worker that was mainly foraging for a few hours in the morning and whose activity was still rhythmic (but with a shorter frequency) in permanent light conditions. ***Bottom right:*** Actogram of a worker that was highly active throughout the daylight hours, and which also displayed a shorter rhythm under permanent light conditions. Shortly before death, this individual shows an increased level of arrhythmic activity ("death dance").

Tagging all workers of many bumble bee colonies, and automatically monitoring each individual's activity, we discovered that under the permanent daylight north of the polar circle in summer, bumble bees still have pronounced daily rhythms, with several hours of rest on a nightly basis. Sleep is clearly important for bees' well-being, even in the brief arctic summers, when colony development must be completed in just a few weeks, bees race against time to find sufficient resources to raise new queens and males, and there should thus be strong pressure to work as hard as possible. But the lifelong monitoring of bumble bees' activity patterns also revealed profound individual variation in temporal foraging patterns. Some individuals worked for weeks throughout the daylight hours, while others favored just the early morning hours, and some completed only a single foraging flight per day. Near the end of their lives, some individuals displayed a "death dance"—a form of arrhythmic hyperactivity without any apparent breaks that has also been observed in fruit flies, perhaps indicating increasing neuronal malfunction and impending death (figure 10.1).

## Same Genes, Different Outcomes: Specialization in Bee Colonies

To scientists working on social insects, it had of course long been known that one of the most spectacular features of social insect colonies is their division of labor, where individuals specialize in one of the many tasks that need to be executed to ensure that the colony as a whole operates like a smoothly oiled machine. An extreme example is found in the dimorphism between workers and queens, and the behavior differences that come with their different "jobs" in the colony.

Honey bee queens and workers are genetically indistinguishable, but what seals their respective fates is that queen larvae get fed a special "designer diet"—the so-called royal jelly—in large quantities and over extended periods. This richly nutritious substance's chemical composition is only partially understood. It is produced by glands in the mouths of young nurse bees. All larvae are initially fed with royal jelly, but worker larvae are soon weaned and switched to a diet of pollen and nectar, whereas queen larvae are bathed in royal jelly throughout their larval development and feed on it into adulthood. This differential rearing procedure results in striking morphological, behavioral, and physiological differences between these different castes (figure 10.2).

Honey bee queens live for years, produce up to 2,000 eggs per day, and never visit flowers (or engage in any other activity of colony construction or maintenance), and their behavioral goals are entirely different from those of worker bees. These goals come with a wholly different psychology: much of the worker bee's mind is occupied with flower visitation, whereas a queen's desiderata are more Shakespearean: upon emergence from their pupae, new honey bee queens engage in a series of deadly duels with rival queens. The single survivor will leave the home for one to five mating flights, during which she visits drone congregating areas used solely for mating, which might be several kilometers from the hive, where hundreds of drones typically wait. Queens will mate with an average of 12 drones in flight; the drones die shortly afterward, since the explosive ejaculation ruptures the everted genitals. A mated queen then returns to her native hive; egg laying begins soon after, and she will typically not leave the colony again unless a new queen is raised in the subsequent year, in which case the old queen leaves the nest with a large swarm of workers to relocate to a new home.

**Figure 10.2. Honey bee queens and workers are genetically indistinguishable, but anatomically, physiologically, and psychologically different.** The honey bee queen (the individual marked with a number tag) and sterile workers display differences related to their different functions in the bee society. Specialized workers who form the queen's retinue feed the queen and constantly groom and lick her, in the process picking up queen mandibular pheromone, which suppresses ovary development in workers.

In stark contrast to a queen's life, the sterile honey bee workers typically live only for weeks, during which they engage in a series of specializations, among them the cleaning of comb cells (the first few days after emerging from the pupa), tending brood or the queen (~days 3–20), constructing wax combs (~days 7–20), guarding the nest entrance (~3 weeks of age), and foraging (typically 2–3 weeks of age) for various commodities such as nectar, pollen, water, and resin. On the other hand, workers will never know sex.

There are also striking differences in the sensory apparatus. Honey bee workers have 60 percent more facets in their compound eyes and 70 percent more olfactory sensors on their antennae, and Maurice Maeterlinck (whose century-old writings we encountered in chapters 1 and 8) commented on the "somewhat empty skull" of the honey bee queen. The many differences in the life span, specializations, behaviors, sensory physiology, and brain anatomy of social insect queens and workers—often solely as a result of the difference in larval rearing—are perhaps one of nature's most extreme examples of the influence of the environment on an individual's fate.

The changes from one task specialization to another within an individual's life span are also reflected in brain anatomy. For example, the transition from within-hive duties to foraging in workers is accompanied by drastic (15–20 percent) enlargements in the mushroom bodies of their brains, presumably as a result of having to memorize large amounts of information about the spatial foraging environment and the features of rewarding flowers. However, some of this growth happens shortly before the age at which bees are destined to leave the hive to forage. This indicates that the bees' inborn developmental programs prepare the brain for outdoor flight by increasing its memory storage capacity.

## Division of Labor as a Result of Individual Differences in the Sensitivity of Senses

The success of insect societies, e.g., ants, bees, and termites, has often been attributed to their labor division, specialization, and the resulting efficiency of the colony. Individuals of insect colonies are indeed often highly specialized, so that animals will predominantly engage in colony defense, nursing larvae, removing debris, or foraging only for particular available commodities but not others.

With the exception of rigid castes, such as egg-laying queens or termite "soldiers," however, specialists are often not distinct in morphology, and indeed are largely totipotent in terms of the tasks they can potentially perform. Even though social insect specialists might perform the same routine for extended periods, with the same repetitiveness as assembly line workers, they can typically switch to other activities should these become necessary. If more "hands" are needed, say, for colony defense, or for foraging, individuals currently engaged with other tasks swiftly abandon those and take on the tasks that are needed more urgently. Early in the nineteenth century, the Swiss naturalist François Huber (see chapter 4) proposed a groundbreaking idea of how this might come about by simple self-organization, without the need for a powerful decision maker allocating workers to one task or another.

Huber was interested in the climate control in honey bee hives, and specifically the question of how they kept the hive well ventilated to avoid suffocation. He noticed that with decreasing oxygen levels, more bees would stand still and whir their wings for ventilation—when the air was extremely stuffy, *all* workers would. Under normal conditions, only a minority of workers would engage in fanning behavior, arrayed in seemingly strategic positions from the hive entrance to the interiors of the hive, without there being any apparent communication between these workers. Huber hypothesized that individual honey bees were differentially sensitive to noxious smells, and that those most sensitive would be the ones to initiate fanning first. Should conditions nonetheless deteriorate, more individuals' tolerance thresholds would be reached, and they would begin fanning too. In this way, a decentralized allocation of the appropriate numbers of workers to the job of ventilation would be assured in all areas of the hive. Huber could not himself test this elegant hypothesis, since his team had no means of marking individual bees. Today, there is ample experimental evidence that the flexible way in which bee colonies allocate workers to the relative urgency of the many vital tasks is indeed mediated, at least in part, by different individual sensitivities to the stimuli that indicate the respective needs.

To explain this form of decentralized labor allocation, the American social insect biologist Jennifer Fewell has used the metaphor of a multiple-person household where, invariably, the same unfortunate person does the dishes. Why? Because each person has a different sensitivity to the stimulus that is the growing pile of dishes in the sink. The person with the highest sensitivity will take on the task first—which removes the stimulus, so the next most sensitive person's threshold will possibly never be reached. That is—unless the prime "dishwashing specialist" is away on vacation, in which case the pile will likely have to grow slightly higher before the person with the next highest sense of tidiness will swing into action. But the job can get done without this necessarily requiring any central organization, or even any collective assessment of need. Importantly, the specialization will be influenced by experience as well: the person whose threshold is reached first also gains the most experience with the job, and will in turn be the one who does it best, which may further cement individual differences in labor specialization (in insect colonies as well).

Individual sensory thresholds were first experimentally determined by Karl von Frisch, as reported in his 155-page opus on the honey bee's sense of taste, which contains a two-page section headlined "Individualität" (individuality). Von Frisch tested bees' readiness to accept low-concentration sugar solutions, or solutions that had been laced with adverse tastants such as hydrochloric acid. He observed individual bees for up to 24 days, and discovered that some bees were uniquely and consistently picky about the minimum sweetness levels they would tolerate, or singularly sensitive to acids or bitter substances. In fact, one individual appeared superlatively sensitive to *all* tastants that von Frisch tried. It was later discovered by the American entomologist Robert Page that differences in sensitivity to sugar are already manifest when bees are just a few hours old, and determine, for example, whether individuals become pollen or nectar foragers weeks later.

## Individual Differences in Body Size, Sensory System, and Work Specialization

Unlike in honey bees, whose workers are all roughly the same size, bumble bee workers inside a single colony can vary drastically in size—by more than a factor of 10—from the smallest, housefly-size workers to some that are practically the size of a queen. Bees don't grow once they have emerged from the pupa, so differences in size of bumble bees of one species that you might see in a colony or on flowers are not related to age. Instead, such variation is the result of differences in the amount of nutrition received during larval development. Bees do all their growing while they are helpless, legless grubs sitting in brood cells.

In adults, there is no *strict* division of labor in accordance with body size in the bumble bee colony, but there is a tendency for the smallest workers to engage more with in-nest duties such as wax construction and brood rearing, whereas large workers tend to be those that leave the nest to visit flowers. My PhD student Johannes Spaethe, working jointly with fellow PhD student Anja Weidenmüller, found out that this makes sense because it turns out that the largest workers in the species *Bombus terrestris* are also the most efficient workers. This is not, however, just the result of physical strength, which might make them better flyers and more efficient at manipulating flowers. It turns out that larger workers also have a superior sensory apparatus.

Johannes discovered that larger workers do not just have larger eyes. Their compound eyes also have larger facets (larger lenses) that convey higher light sensitivity, and this allows them to forage in dimmer ambient light conditions—for example, early in the morning before sunrise, when most other pollinators might still be sound asleep. In addition, by means of a sophisticated technique for shining light beams through the optical apparatus of bumble bee eyes, Johannes discovered that larger bumble bees also enjoy the advantage of seeing higher-resolution images—they see more "pixels" (large workers can have more than 4,000 ommatidia, while small ones may have fewer than 3,000), and the pixels are of a smaller size, than those of small workers, which have pretty blurry vision. This allows large workers to detect smaller flowers, and from a greater distance. In fact, because larger bees carry bigger, higher-resolution eyes, a 33 percent increase in body size is accompanied by doubled precision in flower detection (figure 10.3).

Johannes also discovered that larger bumble bee workers have a keener sense of olfaction: their antennae have a higher number, and indeed a higher density, of olfactory sensors. Pore plates

**Figure 10.3. Larger bumble bee workers have larger eyes with higher light sensitivity and higher resolution.** Scanning electron micrograph of the compound eye of a small (***left***) and a large (***right***) bumble bee (*Bombus terrestris*) worker. Insets show the size differences in facets in the central part of the corresponding eye. Larger workers have larger eyes with higher sensitivity and better resolution, conferring better ability at detecting flowers from a distance. Single scale bar, 50 μm; double scale bar, 500 μm, for both the left and right images.

(the most abundant type of these sensors) range in numbers from ~700 in the smallest workers to ~3,500 in the largest ones, and their densities from ~2,400/mm$^2$ to ~3,200/mm$^2$, which means that large workers can detect floral scents from substantially greater distances. In other words, the (at least partially random) processes that lead to some larvae having better access to food result in pronounced differences in how the adults perceive the world, and determine their later work specialization.

## Task Specialization as a Result of Experience

In social bees, just as in human societies, the choice of "profession," or indeed efficiency at a particular task, is only partially a result of innate predisposition, as determined by sensory thresholds, "talent," or innate tendency to engage in a job. It is also a result of perfecting skills through experience. There is extensive evidence that learning is involved in almost any task performed by social insect workers, including food type recognition and handling techniques (such as for flowers; see chapter 7), but also in such seemingly instinct-driven tasks as nest building (see chapter 4). Direct evidence that early experience of success at a task might to some extent determine the "profession" that an insect worker chooses in later life comes from raider ants (which attack other ant nests to feast on their brood).

The workers of the ant species *Ooceraea biroi* are clonal—genetically identical—and therefore any differences in labor specialization can only be the result of environmental factors. In one study, previously naïve ants repeatedly explored their environment for food—only the experimenters had made sure that some individuals never found any. Such ants gradually decreased their efforts, and in the end, stayed mostly in the confines of the nest and became specialist brood carers, whereas their more successful (genetically identical) relatives continued to forage in the outside world. In this case, the experience of success and failure determined specialization.

Thus, in social insects as in humans, task specialization can also be the result of a self-assessment of whether or not one is successful at a certain task. Contrary to humans, however, there is likely no feedback from others about task performance: no bee says to another, "Hey Jane, you totally *suck* at foraging!" In bees, we don't yet have direct evidence that personally experienced success at a certain task determines the job an individual takes on in the colony longer-term, but the ant study certainly makes this a possibility worth exploring.

## Individually Different Foraging Routes

In 1994, I began working at the State University of New York at Stony Brook as a postdoctoral fellow. My mentor James Thomson had extensively observed individually labeled bumble bees visiting wildflowers, had appreciated the profound variation in their foraging behavior, and thus encouraged my further exploration of individual differences in the behavior of bees. We shared an interest in the "traplines" formed by bumble bees—where bees visit a series of flowers (or flower patches) in (somewhat) stable sequence (see chapter 6). When different bees are left to explore and form such traplines on their own under completely identical conditions, each bee will arrive at its own individual signature of solving the problem, with no two bees showing exactly the same pattern (figure 10.4).

My team explored such individual signatures under field conditions later, when we followed the entire foraging careers of individual bumble bees with radar, from their maiden flight through their discovery and exploitation of flower resources to their death. In chapter 6, we have already encountered an individual that, after two early exploration flights, only ever visited two foraging locations over her entire life. However, not all individuals' records show them to be this faithful to particular foraging locations. One radar-tracked bumble bee never settled on a single foraging patch during her life: almost every one of her foraging bouts was exploratory in nature, with relatively little dedicated resource exploitation (figure 10.4, bottom right), even though plenty of rich flower patches were available, and other bees returned to them regularly. It is doubtful that this individual ever contributed much to the

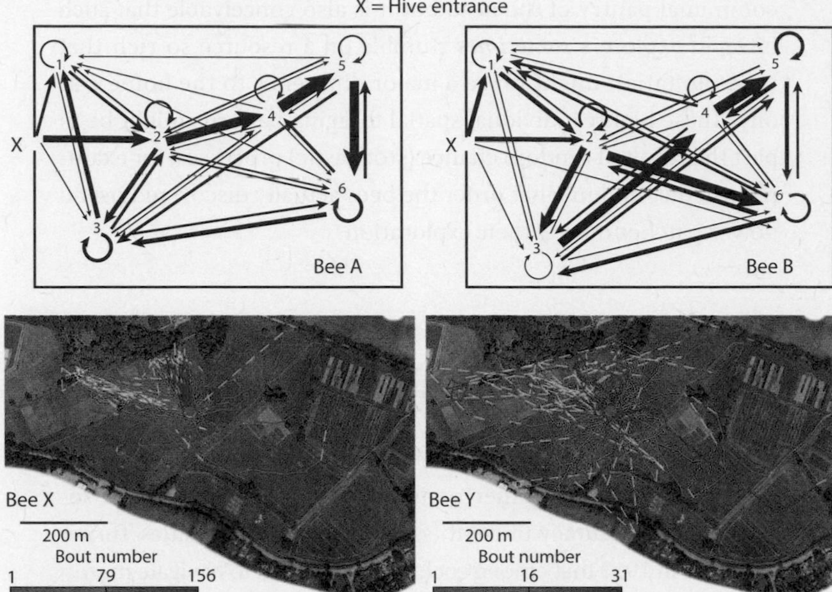


**Figure 10.4. Individuality in spatial foraging strategies of worker bumble bees in a laboratory flight arena (top) and under natural foraging conditions (bottom).** ***Top:*** Examples of routes taken by bees foraging on six artificial flowers (whose positions are numbered 1–6). The outline of the flight arena (105 cm × 75 cm) is marked with a thin rectangle. The width of the arrow corresponds to the frequency with which each trajectory was taken throughout 40 sequential foraging flights. Circular arrows: cases when a bee revisited a flower just visited. Though the two bees faced exactly the same foraging situation, they had highly individualized favored sequences: e.g., bee A tended to move in a path straight ahead from the hive entrance and then turned right when the far wall was reached. Bee B strongly favored routes from flower 3 to 4 and 2 to 6, which were rarely taken by bee A. ***Bottom:*** Flight paths of every flight made by two bees over their entire lives under free-flight conditions in the same summer. Blue circle: position of nest. Earliest flights by each bee in green, changing through yellow until the last flights in each bee's life are shown in red.

communal pantry of the nest, but it is also conceivable that such intrepid explorers *sometimes* stumble on a resource so rich that its exploitation might make a major difference to the home colony. These highly individual spatial foraging patterns might be in part the result of random chance (stochastic) processes, for example whether and in what order the bees actually discovered useful flower patches during their exploration.

## Individuality in Speed-Accuracy Tradeoffs

One psychological trait in which individual variation was observed in insects before any other nonhuman animal concerns the so-called speed-accuracy tradeoff (see also chapter 7). Charles Turner observed in 1913 that among cockroaches trained to navigate mazes, younger individuals tended to be fast and error-prone, whereas older ones were slower but made fewer errors. Generally speaking, in any difficult discrimination task (such as telling apart two similar colors, patterns, or numbers), one can place emphasis on accuracy, but this may take an extended inspection time, or on speed—in which case accuracy may suffer. In bumble bees, we found no age differences in this regard, but we did find that there are differences between individual bees in how they go about this problem: some bees are consistently fast and sloppy, whereas others are more careful, slow, and accurate in their decision making. Since individuals with different preferences for speed or accuracy might fare better or worse under different ecological conditions, the colony as a whole might fare best by harboring a diversity of individuals with different strategies (figure 10.5).

Such speed-accuracy tradeoffs in bees have been discovered not just in the discrimination of flowers' colors, but also in predator

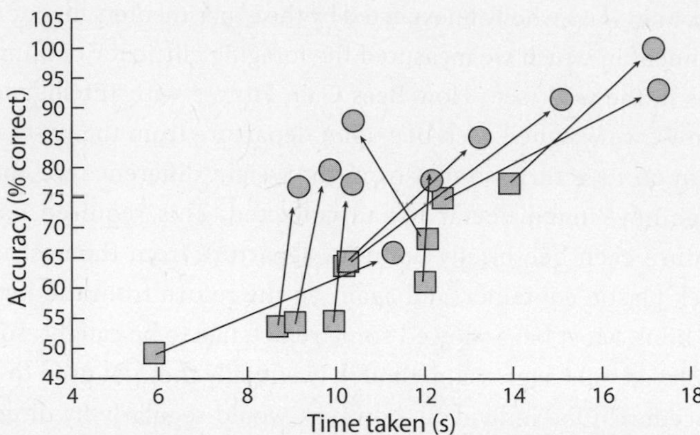


**Figure 10.5. Bumble bees can choose wisely or rapidly, but not both at once.** Interindividual correlation between response time and accuracy of bees discriminating between two similar flower colors. Each symbol denotes the average performance of one individual bee under one experimental condition. When targets were rewarded with sucrose solution and alternative colors contained no reward—i.e., plain water (square symbols)—bees investing more time made more accurate choices. When choosing alternative colors was penalized with bitter quinine solution (circles), all bees improved their accuracy. Arrows link the average values for individual bees under the two experimental conditions.

detection. Invariably, different individuals place their priorities in different ways on either speed or accuracy. Such diversity among the workers may well benefit colonies as a whole.

## Individual Differences in Intelligence

When one performs experiments on the learning behavior of bees, there are often one or two "genius individuals" that solve a problem more quickly than all others, or in an exceptionally efficient way, or in ways wholly unexpected by the experimenters. In one experiment in which we measured the foraging efficiency of bumble bees in the wild (see "How Bees Gain Fitness with Intelligence," below), we weighed each bee upon departure from the nest, and again on its return, so that from the weight difference we could judge how much nectar it had collected. This required us to capture each bee briefly upon its departure from the nest in a black plastic container, and again on the return from the foraging bout. Most bees showed some reluctance to be caught; some displayed mild aggression, though eventually they got used to the procedure. One individual, however, would regularly fly directly into the black container, even if an experimenter held the container overhead meters away from the hive: this bee had essentially come to view the container as a "public transport" vehicle and expected to be carried back to the nest inside it. The individuals that are exceptionally innovative at problem solving are typically those whose behavior is the most variable, and which thus appear more exploratory than others. In this way, intelligence is linked to behavioral variability: the German neuroscientist Björn Brembs builds a convincing case that fully hardwired, predictable behavior is a sure path to extinction. For example, if an animal species behaves in a fully foreseeable manner when confronted with a predator, the predator will figure this out eventually, and it will be "game over" for the prey. Having some—though not unlimited!—noise in the nervous system means that behavior always has some level of variability. Those individuals with more strongly pronounced behavioral variability will experiment with more solutions to a problem, and will thus ultimately be more efficient problem solvers.

This became apparent in our experiments with string-pulling bumble bees, in which bees had to pull a thread to get access to an artificial flower placed under a plexiglass table (see chapter 8, figure 8.3). The vast majority of individual bees (over 100 in this case) either required stepwise training, or had to observe other bees solving the task before they managed it themselves. Two individuals, however, solved the task spontaneously, and it was clear from our video recordings that these were especially exploratory individuals, who tried tirelessly to reach under the plexiglass table from a variety of positions, using various body postures, until their feet caught hold of the string, causing a visible flower movement that prompted them to elaborate on the technique.

In other experiments, which do not require a specific innovation or insight from the subjects, the differences between individuals are more of a gradation; they are quantitative rather than qualitative. In such tests it is possible to assign numerical values to individuals' performances, for example by quantifying and comparing learning speed in the same task (such as learning that one artificial flower type is rewarding and another is not). By following each individual bee's learning progress over time, and measuring how it improves with experience, one can use mathematical tools to fit curves to each bee's learning behavior (chapter 7, figure 7.2). This allows a rigorous quantification of the steepness of each individual's learning curve. Note that when people refer to a "steep learning curve" in everyday life, they typically mean a challenge is demanding. However, in reality it's just the opposite: a steep learning curve means rapid improvement of performance—thus an easy-to-learn task or an exceptionally astute subject—whereas a gentle slope of the curve means that performance improves only gradually, indicating that the task is hard, or that the individual isn't a particularly fast learner.

Such tests reveal that there is pronounced variation between individuals' learning performances (see figure 10.6). Moreover, individuals that are good on one particular task, such as learning about the colors of rewarding flowers, also tend to be better at learning visual pattern discrimination and learning to distinguish between

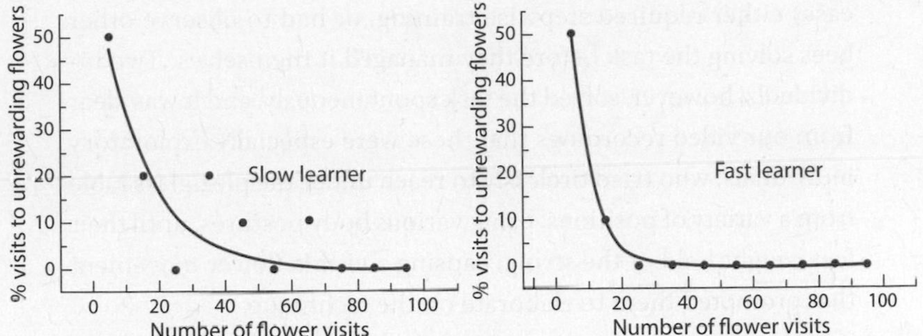


**Figure 10.6. Learning performance differs strongly between two individual bumble bees.** In this experiment, bees had to learn that yellow-colored artificial flowers are rewarding and blue ones are not. Both bees start out visiting rewarding and unrewarding flowers at 50% each. As they gain experience, both bees reduce visits to the unrewarding color to zero, but at individually different speeds.

floral odors. The observation that individuals who are especially clever at one of type of cognitive task often tend also to be good at others is of course familiar from humans, leading some psychologists to believe that a single factor determines abilities in a wide variety of tasks. This is commonly referred to as *domain-general learning*, and the correlation between abilities in different tasks is measured as the factor G (general intelligence). Such measures have been explored in a variety of animals, but not yet in bees.

## Oskar Vogt: from Bumble Bees to Lenin's Brain

If there are differences in individuals' psychology and intelligence, then these must somehow be underpinned by differences in the brain—not necessarily in its overall structure, but in its internal wiring diagram. The question of which properties of the brain determine an individual's intelligence has fascinated scientists for well over a century, though of course mostly in the context of humans. But few people know that one of the pioneers in researching such differences in humans, Oskar Vogt (1870–1959), drew his inspiration to study individual brain variation from his childhood observations of bumble bees. While still in high school he had observed variation in the body coloring of individual bumble bees of the same species, and, having studied Darwin's teachings on individual variation and selection, he thought that such variation must be relevant to evolution, whether it occurred in bumble bee coat colors or in human brains.

Vogt was an internationally renowned German neuroscientist, and his interest in the individual variation of brain anatomy in humans led him to search for the neuroanatomical correlates of genius. Because of this interest and renown, he was invited to Moscow in 1924 to examine Lenin's brain after the Soviet leader died of a stroke. Vogt prepared over 30,000 sections of the brain, and, finding that Lenin had particularly large nerve cells in certain cortical layers, declared him a "brain athlete and association giant." This somewhat opportunistic verdict was no doubt announced to please his Soviet sponsors. It does little justice to the pioneering and otherwise rigorous work of Oskar Vogt and his wife Cécile Vogt-Mugnier (1875–1962) who, for their breakthroughs in understanding the cellular architecture of the brain, received 13 joint nominations for the Nobel Prize.

They were never awarded the prize, and because of their scientific links to the Soviet Union, left-leaning political views, and sympathy toward Jewish scientists, they became an immediate target of the Nazis after the party took power in 1933. Following multiple brutal raids by Nazi storm troopers ("SA") on his laboratories and his private home, Oskar Vogt received a personal letter from Adolf Hitler in 1935, announcing that his retirement had been decreed. Vogt was forced to hand over his directorship of the world-famous Kaiser Wilhelm Institute for Brain Research, of which he had been a founder, to Hugo Spatz, a Nazi Party member who would perform much of his work on the brains of victims of the Nazi euthanasia program.

## Individual Differences in Brain Structure and Intelligence

The Vogts subsequently moved to the Black Forest and continued to work on a privately funded basis. They continued to publish on the relationship between variation in the human brain and the more easily observable heritable variations in animals, including bumble bees. Unfortunately, they never thought of closing the loop—of using insects as a more accessible model to find the neural substrates of individual intelligence, as had been attempted with Lenin's brain. So my team did it for them, exploring whether the individual variation in bees' color learning speed could be explained by differences in brain structure.

Color learning ability can be linked to natural foraging success—see "How Bees Gain Fitness with Intelligence" later in this chapter—and also correlates with other measures of learning. But just measuring the size of nerve cells, as Vogt did in Lenin's brain, says little about the capacity for associative learning, which is mediated by changing the connections between nerve cells, the synapses. As we learned in chapter 9, the mushroom bodies are the principal association centers of the bee brain. Many axons from the visual centers (optic lobes) of the bee brain terminate in the mushroom bodies, where they form connections with the mushroom bodies' intrinsic cells, the Kenyon cells. In the same input region of the mushroom bodies (the so-called "collar" region) where

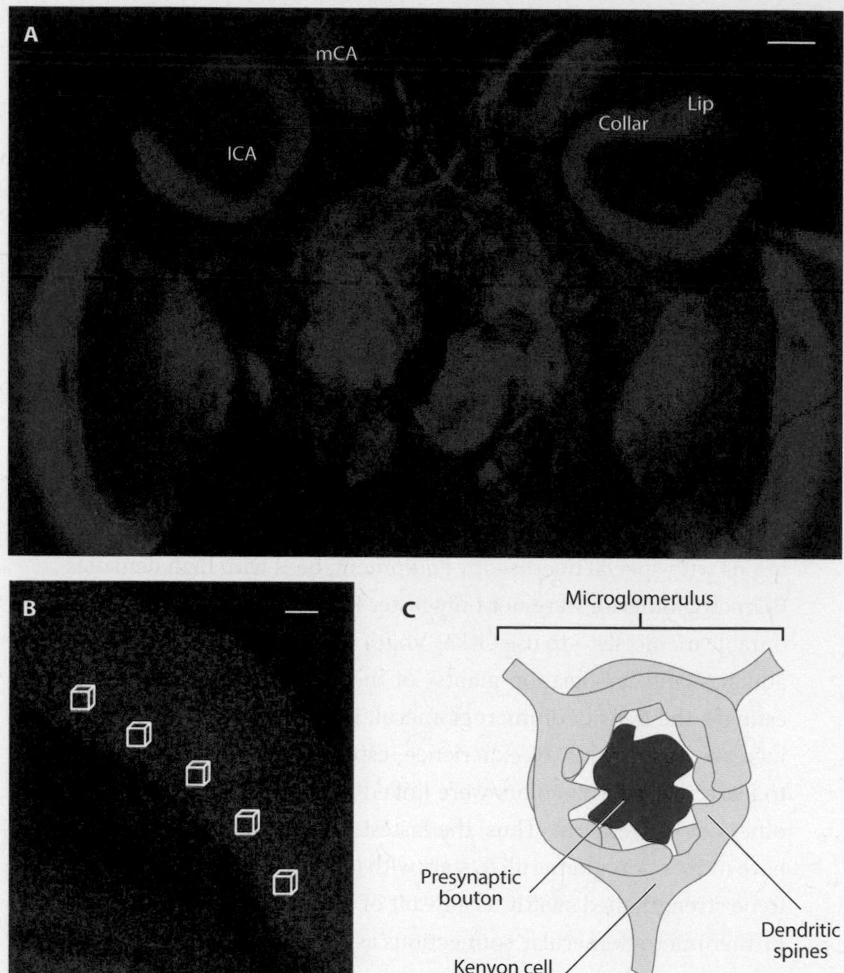


**Figure 10.7. Density of connection points in the bumble bee brain determines learning ability.** ***A.*** Frontal view of a bumble bee brain in which synapses (connection points between neurons) are colored red (scale bar: 150 μm; lCA, lateral calyx; mCA, medial calyx). ***B.*** Collar region of a mushroom body (scale bar: 20 μm): individual microglomeruli (synaptic complexes) can be seen as red dots. White outlines delineate example positions of selected cubes in which we counted microglomeruli—more such glomeruli correlate with better learning. ***C.*** Diagram of a microglomerulus, including a presynaptic bouton (terminal of the axon) of a sensory nerve cell (red) and the input region (dendrites) of Kenyon cell neurons (grey).

visual information is processed, there are also endings of the neural reward pathway, which signals when a sweet reward is perceived by the bee's mouthparts. The connections between the sensory inputs (visual information and reward signals) and the Kenyon cells are synaptic complexes called microglomeruli (figure 10.7). These connections are plastic—meaning that both their number and connection strength can be modified by learning if visual information coincides with reward. It thus makes sense to assume that individual bees with higher densities of microglomeruli in the mushroom bodies' collars might be better learners, since they have more connection points that can be strengthened by learning.

This is indeed what we found by looking deep into the bees' brains with special microscopy equipment: bees with high densities of microglomeruli were not only faster learners, but also had more durable memories—to use Oskar Vogt's words, they were the "brain athletes" and "association giants" of the bumble bee world. Interestingly, the density of microglomeruli in this brain region further increased as a result of experience, especially when the bees had to learn that several colors were linked to reward, whereas several other colors were not. Thus, the fastest learners may be those that have more microglomeruli to start with (allowing more connections to be strengthened swiftly as a result of experience) and then build further microglomerular connections as experience accumulates.

## How Bees Gain Fitness with Intelligence

Just as there are "personality" differences between individual bees, there can also be such differences between colonies of social bees. Beekeepers know well, for example, that some hives are uniquely aggressive, while others may be particularly good honey producers.

Such differences are unsurprising: bee colonies—whether of domesticated honey bees or wild bees such as bumble bees—are families of highly related individuals that share many of the genetic factors that determine their behavior. Even though, as we have seen in the preceding paragraphs, individuals *within* a single colony may display pronounced behavior differences, there are still greater such differences *between* colonies. Each colony has its own behavioral signature that distinguishes it from other colonies of the same species. This concerns psychological factors like aggression, and also various aspects of cognition, such as learning speed.

In the 1980s, Christian Brandes in Randolf Menzel's team managed to generate honey bee selection lines of superior and inferior learners—by selectively cross-breeding the offspring of colonies whose workers learned rapidly with those from colonies of poor learners. This was a direct demonstration that learning performance in bees has a genetic basis, is heritable, and can be subject to selection. And if selection results in changes in learning behavior over just a few generations in controlled laboratory conditions, this means that selection can also act under the often much more stringent conditions in the wild. Natural selection does not tolerate failure—not in running from predators, nor in coping with disease, nor in quickly processing salient information. To be slow on the uptake is as disadvantageous as being slow of wing or foot.

Thus it makes intuitive sense that faster learners should do better in the wild—but how much better? Until the mid-noughties, we still knew almost nothing about how learning performance is adapted to the real ecological conditions in which animals operate. We thus wished to explore whether in bees we might demonstrate a direct link between variation in learning ability and foraging performance. We tested a large number of individual worker bees of twelve bumble bee colonies in a flower color learning task where one color was linked to sugary rewards and the other was not.

Learning curves (see figure 10.6) were measured for each bee under controlled laboratory conditions, and once we had tested enough bees from each colony, we then placed the same colonies in the open, so that they faced the real-life challenges of locating and learning about suitable flowers in the colonies' large flight ranges. We weighed each individual upon departure from the nest and upon its return, so we would have the trip duration as well as the net weight gain of the nectar collected.

The results were striking. Colonies varied in learning speed by a factor of nearly five, and the colonies dominated by the slowest learners collected 40 percent less nectar than the colonies containing, on average, faster learners. This indicates that high learning speed might confer substantial advantages under natural conditions. On the other hand, even members of the slowest-learning colony didn't come home entirely empty-handed, suggesting that the most rapid learners don't deplete *all* the goods.

## Why Aren't Slow Learners Extinct Yet?

If natural selection favors faster learners, why are there any slow learners left in the wild at all? Are there some disadvantages to making associations rapidly that might allow slow learners to persist under natural conditions for many generations?

We explored this question from many angles. For example, we wondered whether rapid learning might lead to such tight associations that it might interfere with the acquisition of new information when previously learned contingencies are reversed, such as when a previously rewarding flower species or patch has been overexploited and is tapped out, and another, previously poorly rewarding species ups its nectar secretion and is now a food bonanza.

But it turned out that those individuals that learned rapidly were also swift at reversing their associations. We also found that bumble bees that were good at learning colors also tended to excel at learning shapes and odors: again, there seemed to be no tradeoff between performance at one task and at another; instead, smart individuals tended to perform well at all tasks.

Taken together, these findings made the persistence of slow learners in the wild an even bigger mystery. If fast learning is strongly advantageous in the wild, and has no costs, why do we still see slow learners at all? One potential clue came from a study in which it was found that faster-learning bumble bee individuals were active for fewer days of their short lifespan than were slow learners, and this effect was so pronounced that over a lifetime, the "dumber" individuals actually contributed more to colony foraging success. Perhaps the reduced foraging activity in the smarter bees was a result of an energetic cost of rapid learning.

To conclude this chapter, we have seen that there are immense differences in sensory systems, behavior, and learning between individual bees and between colonies. Viewing bees as beings with unique "personalities," possessing individual preferences, learning abilities, and memories also lends a new perspective to the need for their conservation. In 2016, we launched the London Pollinator Project—an initiative to encourage Londoners (and ideally inhabitants of other cities) to plant more pollinator-friendly flowers, such as English lavender (*Lavandula angustifolia*), viper's bugloss (*Echium vulgare*), or spiked speedwell (*Veronica spicata*). Planting such flowers is exceptionally helpful to supply wild pollinators with the nutrition that has become scarce as a result of urban sprawl, industrialized agriculture, and the fact that many gardeners plant specially bred flowers that are large, showy, and pleasing to the human eye, but entirely useless as food sources to bees.

As a symbolic link between us scientists and the community, we marked over 2,000 bees of three species individually with two- and three-digit number tags in multiple colors. The bees' nests were based on East London's Queen Mary University campus, and the labeled bees could forage freely in gardens, parks, and balconies across London. People could thus see individual bees return to particular gardens repeatedly, and sightings of labeled bees were reported from sites as far as eight kilometers away from the hives. The idea behind this project was that observing number-tagged bees in their gardens would raise people's appreciation of bees as individuals—with unique biographies and memories of particular flower patches, and with individual flower preferences that differ from those of other bees. Once you view animals as individuals rather than anonymous entities, you develop a connection with them, and a deeper understanding of why it is important to assist in the conservation of threatened animals.

The results were encouraging. The project received a huge press echo, and our interactive web page logged many comments from London's citizens, indicating that they now understood pollinators not just as an anonymous commodity, worth conserving because we need them to pollinate our crop plants, but also as individual beings with unique life stories. Many said they felt a sense of regret when a familiar bee ceased to come to their garden at the end of its relatively brief life. Perhaps readers will come to appreciate the need for their conservation even more in the coming chapter, where we will learn about bees' "inner lives" and explore the question of whether they feel and subjectively experience the world around them—whether they have a form of conscious awareness.

# 11

# Do Bees Have Consciousness?

> The mother-bee produces ten thousand individuals at one time; if these ten thousand individuals were still more stupid than I believe them to be, they would still be compelled to organize themselves in some fashion, in order to continue their existence. . . . Place together, in the same room, ten thousand automatons animated with a living force, and all induced, through the perfect resemblance of their outer and inner being . . . if we admit the least degree of feeling in these automatons, even only such as is necessary for them to be conscious of their own existence, seek their own conservation, avoid noxious things, prepare useful things, &c, their work will not only be regular, well proportioned, similar, equal, but it will also have symmetry, strength, convenience to the highest point of perfection.
>
> **—Charles Bonnet, 1764**

Do bees have subjective experiences, feelings such as pain, and are they "conscious of their own existence"? We have learned in chapters 2 and 3 that, in a sense, all experiences are subjective—in that sense organs never send an "objective," veridical reflection of the world to the brain, but always one that is filtered by the sensors that have been acquired over evolutionary time to suit the needs of the particular animal. This means, for example, that a poppy flower that has a reflectance curve with peaks at electromagnetic radiation wavelengths below 380 nm and above 600 nm (but little reflectance in between) looks red to us, but to a bee looks completely different, since it perceives the UV but not the red reflectance. We can experimentally ascertain that bees see this UV reflectance and we do not. But *how* bees actually perceive it—how it *looks* to a bee, subjectively—is fundamentally unknowable. And so it is with all subjective experiences.

We therefore have to content ourselves with common sense and probabilities. When you see someone cry, even a total stranger, there is good reason to believe that they have been subjected to an upsetting emotional experience (you don't know this for sure—they might be faking it—but you can be *reasonably* sure). When neuroscientists found that rats "replay" the same sequential brain cell activation patterns at night as they showed the previous day while learning a maze, it is justifiable to conjecture that they "relive" their memories in their sleep. Having access to autobiographical memories in the absence of external triggers is a hallmark of consciousness. When it comes to the experience of pain, we have little doubt that when a dog winces and yelps at a foot injury, limps, and protects the injured foot, it experiences something more than a simple, suffering-free, reflex withdrawal from the source of damage—that the injury *feels* bad to the dog.

In what follows, we will explore the possibility of such experiences and of consciousness in bees. We are of course on speculative territory, but this is a necessary endeavor, at the forefront of science. My postdoctoral mentor James Thomson, commenting on the virtues of uninhibited speculation, once quoted an old Jesse

Winchester tune: "If we're treading on thin ice, then we might as well dance." And right he was. If John Lubbock, for example, had been afraid of ridicule when he experimented with the telephone to explore the ants' language, he would never have discovered pheromone communication (see chapter 3).

Before we start, let's be clear that no one is suggesting that bees' consciousness is in any way as rich and detailed as humans'. I am not proposing that bees ponder the arc of their life from youth to death, that they analyze their own emotional states—"I'm feeling a little down today, I don't think I'll go out foraging"—or that they guess at what is on another bee's mind. But they might have an awareness of the things and living beings around themselves; they might be able to look into at least the immediate future (and plan accordingly); they might experience some form of emotions, and make a basic distinction between "self" and "other."

## Do Bees Feel Pain?

Karl von Frisch thought that bees lack not just the subjective experience of pain, but even any reflex-like response to injuries as severe as cutting off the entire abdomen, if the amputation happens while they are imbibing sugar water. He rationalized that they don't need such responses because they are equipped with an external skeleton. It may be a convenient delusion for scientists to assume that their experimental animals feel nothing of the often invasive procedures to which they are subjected in the laboratory—but it is just that: a delusion.

In analyzing animals' responses to damaging stimuli, one must distinguish between *basic nociception* and *pain perception* (more on the latter below). Nociception is sensitivity to strong mechanosensory stimuli indicative of tissue damage (or the threat of it). Von Frisch denied bees (and other animals with exoskeletons) the basic capability of nociception. That this claim is preposterous is apparent to anyone who has ever witnessed a grasshopper—or an earthworm, for that matter—being impaled on a fishing hook: they struggle with all the vigor that a human might if subjected to the same treatment. It is now clear that many invertebrates (and certainly all insects) have specialized sensory mechanisms to register tissue damage, and segregated neural pathways for nociception and regular mechanoreception.

To respond adequately to damaging stimuli, one needs to have nociceptors in the location of the damage. If a bee fails to respond to the amputation of her abdomen, then the reason may well be that there are no appropriate receptors in the region of the incision. You can economize on nociceptors where you don't need them, or where there is little you can do about tissue damage. The reason human patients sometimes sense very little pain even when they are affected by substantial-sized tumors is that we have relatively few nociceptors in many areas inside our bodies. Before the advent of modern medical procedures, the kinds of threats to which we can generate a meaningful behavioral response tended to come from outside, not inside, the body. The probability of a natural attack on a bee's waistline is relatively slim.

It is also possible that there is a modulation of nociceptive signals at the time of feeding, especially when one considers that the ad libitum feeders used by von Frisch typically contain many thousands of times the reward value of a natural flower (the equivalent of a major lottery win in the human world). The discovery of such an off-the-scale food bonanza might well put the bees into a non-natural state of "euphoria" (see below for emotional state changes induced by even minor rewards) that overrides sensory signals of body damage. If you are nonetheless confident that a bee does not sense potentially damaging stimuli, try to roll one around between your thumb and index finger, even gently, and see if she doesn't swiftly respond in a manner that indicates that she is not pleased by the sensation (as will become apparent through your own nociception).

Contrary to von Frisch's assertion, nociception is essential for survival (figure 11.1) even if you are privileged to possess natural armor (an exoskeleton); and therefore most if not all animals, including insects, have it in some form. Like vertebrates, insects also exhibit wound-healing processes that will be facilitated if the site of the injury is protected for the duration of the repair. We will shortly see that bees learn from noxious stimuli, and exhibit long-lasting behavioral and psychological changes as a result of simulated predator attacks. But is it conceivable that, in bees and other insects, nociception happens without pain, without the subjective experience of suffering?

Pain is different from "simple" nociception in that pain is a subjective, unpleasant sensation whose link with nociception can be modulated by context, attention, and past experience. This flexible link between nociception and pain is apparent in our own experiences. You might come back from a beautiful summer mountain hike, and someone points at your knee and says "Oooh, that's a nasty scrape." You hadn't even noticed, but now that someone has directed your attention to the wound, it suddenly starts hurting. There are reports of soldiers with even severe battle injuries who only begin noticing the pain after returning to safety. In the meantime, their endogenous opiates have kept the pain in check; the body knows how to self-administer such substances while the top priority is escape from the imminent threat of further injury. It is thus clear that a rigid, reflex-like nociceptive system would be of little use to most animals. A biologically useful system for escaping and learning from serious threat and injury would include

**Figure 11.1. An anxiety-inducing situation for a bee?** Bees often come under predator attacks, such as from orb-weaving spiders. Nociception is critical to initiate both escape responses and defensive behavior (biting, stinging). Bees often, though not always, escape such attacks, and if so, have an opportunity to learn to avoid the cues associated with predation threat. New research indicates that stimuli associated with predation can induce emotional "anxiety-like" states in bees.

the possibility of *modulating* the sensation and intensity of suffering according to context.

## The Subjective Dimension of Pain

Because of this subjective nature of pain, it is impossible to measure suffering objectively, or even to assess it in anyone but yourself. Just as in the above example of a dog's foot injury, where the animal's observable behavior gives us reason to believe that it suffers, such reasoning by analogy must in principle also work in animals such as insects that don't vocalize when injured, or display body postures that to a human would betray suffering. But insects cannot express suffering in mammalian body language. Hence we need to turn to physiological and psychological indicators of pain. For example, we have also established that one of the hallmarks of pain perception is that it can be modulated—and we can, of course, measure whether animals' response to adverse mechanosensory stimuli or injury can be controlled according to context.

Evidence of modulation of a pain response exists in honey bees. Because of the immense nutritional benefits of honey (and also the protein- and fat-rich brood of a honey bee colony), many animals, such as bears, rodents, badgers, and skunks, raid honey bee nests. Most animals' natural response to a large predator attack is to flee the attacker (and the potential unpleasant sensations of being eaten), but if you have a home to protect, just running away isn't an option—you have to fight back. Obviously, honey bees are well equipped to mount a counterattack, given their poison gland and an inbuilt syringe to inject the painful poison into the enemy. Honey bees don't sting only when they feel "personally" threatened by a potentially damaging stimulus (such as a spider attack), but also during a socially orchestrated preemptive strike when a threat (say, a large looming shape, like a bear's) appears near the nest entrance. In such a situation, guard bees at the entrance release an alarm pheromone, a scent that sends a signal to recruit a large number of workers to attack the intruder.

But this pheromone does not just make guard bees more aggressive; it also appears to make them insensitive to bodily harm. And this might well be essential for a successful defense against a bear: the attacking honey bees will sacrifice their own lives for the good of their colony. Their stinging device is a masterpiece of bioengineering: it has barbs that make the weapon stick inside the intruder's skin—which means that even if the bear manages to wipe the attacking bee out of its fur, the stinger stays behind. But it's not just the stinger: the poison gland, too, remains attached to it, as does the nerve center that controls its contractions, so that the gland keeps pumping a cocktail of pain-inducing chemicals into the skin of the attacker.

The ripping out of such a substantial organ of the bee's abdomen spells its death—preceded by a potentially highly noxious stimulus that most animals under most circumstances would seek to avoid. The honey bees' alarm pheromone, however, appears to flood their system with an endogenous painkiller that makes them oblivious to their battle injuries. The Argentinean bee scientist Josué Núñez and his team have shown that the more of the alarm pheromone component IPA bees are exposed to, the less they respond to electric shock, until at high dosages, the majority of bees do not respond at all. The result is that IPA appears to reverse the normal escape-survival response, so that guard bees become fearless suicidal attackers.

It is not yet clear, however, what the chemical nature of the painkiller is: the endogenous opiate system, which appears to exist in all vertebrates to modulate pain sensations, is absent in invertebrates. Nonetheless, opiates (and opiate antagonists) do have documented effects in insects—perhaps they simply bind to a non-opiate receptor to generate these effects. It is possible that an alternative endogenous system, perhaps based on the somewhat similar allatostatins (a class of neural hormones that does exist in invertebrates) plays this role in insects. But whatever the nature of the chemical and its receptors, it is thus clear that honey bees have more than a reflex-like response to damaging or damage-threatening stimuli. The fact that such responses can be modulated to the animal's advantage according to the current situation is one of the hallmarks of pain perception.

## Long-Lasting Psychological Changes after Encounters with Predators

Pollinating insects face predation not just in and near their homes, but also on flowers. Crab spiders are sit-and-wait predators that can, chameleon-style, adopt the color of the flowers on which they hunt unsuspecting flower visitors. But crab spiders and bees are pretty well matched for strength and speed—in the majority of cases, bees manage to evade the spiders before the hunters can penetrate their prey's cuticula with their poisonous fangs. But even a bee equipped with a hardwired response might simply seek to escape the mechanical stimulus of the spider's grasp and then continue visiting flowers. A slightly more sophisticated, but almost equally useless, response would be for bees that survived a spider attack to learn to avoid the flower species on which they were attacked. But bees don't have the luxury of abandoning valuable food sources altogether. It is thus clear that an adaptive response would be more flexible, allowing bees to continue visiting the flowers while also minimizing predation risk.

Tom Ings explored the psychological impacts of crab spider attacks on bees in his postdoctoral work on my team, and discovered that attacked bees exhibit sophisticated, long-term behavioral changes that are in line with the predicted psychological effects of a subjective unpleasant experience. We constructed a sort of robotic crab spider—a life-size spider model equipped with sponge-padded electromagnetic pincers that could capture a bumble bee for a brief period (two seconds). Like real crab spiders, the models were the same color as the flowers on which they lurked (although for comparison, some other model spiders were mismatched in color to the flowers). These robotic crab spiders were placed on a fraction

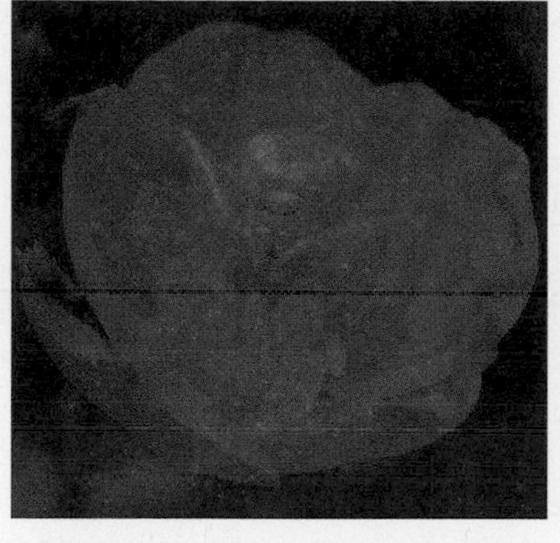

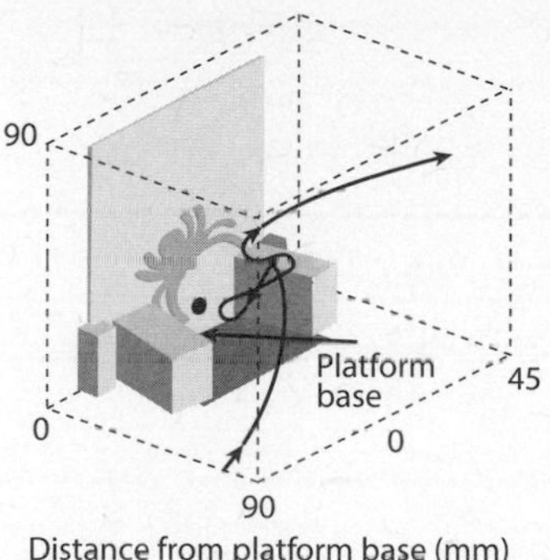


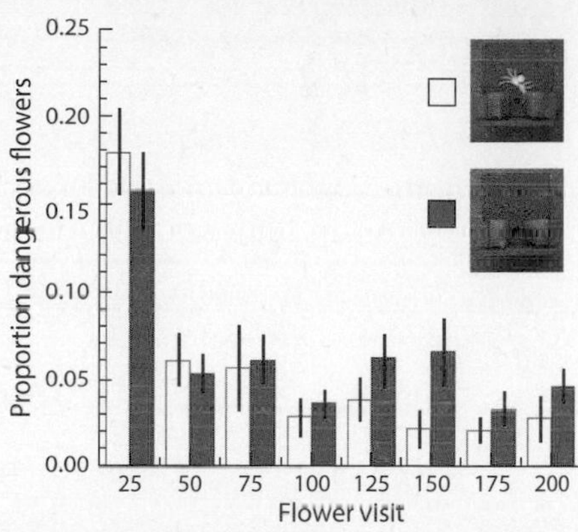


**Figure 11.2. How bumble bees learn about predation threat from crab spiders.** ***Top:*** Crab spiders can change color to camouflage themselves on the flowers on which they wait for unsuspecting pollinators. Bees' responses can be measured with "robotic" crab spiders in the laboratory (***bottom left***). The light grey square is an artificial flower (black dot: feeding hole with nectar) that has a life-size model of a crab spider attached. Above the platform base are the two solenoid-driven, sponge-padded jaws of a pincer mechanism that can capture a bee without harming it (block shapes on either side of the feeding hole under the spider). Bees experienced with such "attacks" subsequently scan all flowers carefully (see black line for a typical bee trajectory). ***Bottom right:*** with an increased number of flower visits, bees manage to avoid landing on spider-infested flowers more and more, though more errors continue to occur when spiders are cryptic (yellow on yellow) than when they are conspicuous (white on yellow).

of the artificial flowers from which bees could collect nectar, so that bees made multiple sequential flower visits, some of which were dangerous and some not.

It was found that bumble bees swiftly learned to avoid the flowers on which they detected a spider (though this took them longer when the spiders were the same color as the flower than when they were easily detectable). But this necessitated a profound behavior change in how bees approached the flowers: they typically spent several seconds inspecting each flower by scanning flight movements. Not only that, but they also displayed "false alarms," in that they sometimes rejected perfectly safe flowers after searching them for hidden threats. Such responses were still apparent 24 hours after training, even under conditions in which there were no spiders at all. The bees' response is thus far more complex than a simple withdrawal from an unpleasant experience: bees learn from the attacks, adapt their behavior to minimize risk of future attacks, and exhibit long-lasting (by a bee's standard) psychological effects, such as "seeing ghosts" of predators even when there are none, indicating an anxiety-like state (figure 11.2).

## Emotional Bees

To explore such emotional states further, the British behavioral biologist Melissa Bateson (whose past work had explored emotions in birds and rodents) teamed up with the American bee scientist Geraldine Wright to adapt the protocols hitherto used to diagnose emotional states in vertebrates to explore their equivalents in bees.

Honey bees were tested with an equivalent of the question "Is this glass half full or half empty?" Humans who are in an anxious or depressed state (or who have anxious personalities) will be more likely to judge the metaphorical glass filled to the midline as half empty—they display a "cognitive bias." Likewise, animals in negative emotional states are more prone to making "pessimistic" judgments about ambiguous stimuli than are those in a positive affective state. Wright and her team trained bees to associate a 9:1 mixture of two odors with a sweet reward, and a 1:9 mixture with a bitter quinine solution which bees dislike. The bees were subsequently presented with ambiguous, intermediate stimuli (such as a 1:1 mixture of both odors). Prior to testing, however, half of the bees were briefly vibrated with a "vortecizer" (a piece of laboratory equipment to mix liquids by shaking), simulating something like a predator attack. The other half of the bees were left undisturbed. The shaken bees appeared to display a "pessimistic" cognitive bias, in that fewer bees of this group than of the undisturbed group accepted the ambiguous stimulus.

In a similar project, my team explored whether a positive emotional state could also be induced in bees faced with a "glass-half-full, glass-half-empty" paradigm (figure 11.3). In this case, a group of bumble bees received a tiny droplet of "surprise reward" before entering the testing arena—and sure enough, such bees displayed a positive cognitive bias, in that they judged ambiguous stimuli as potentially more rewarding than did control bees that had not received the surprise. It is now clear that emotions are survival-related (perhaps survival-critical) states that are not necessarily computationally complex and certainly do not require a large brain. Natural selection might not look kindly upon individuals that do not know fear, mothers who are indifferent to the loss of their offspring, or social animals for whom it does not "feel rewarding" to be in their social setting. In other words, having at least a range of basic emotions might be part of most animals' "survival tool kit."

Another intriguing line of investigation in the context of insect emotions comes from observations that insects appear to seek out psychoactive substances that, in humans at least, are well known to

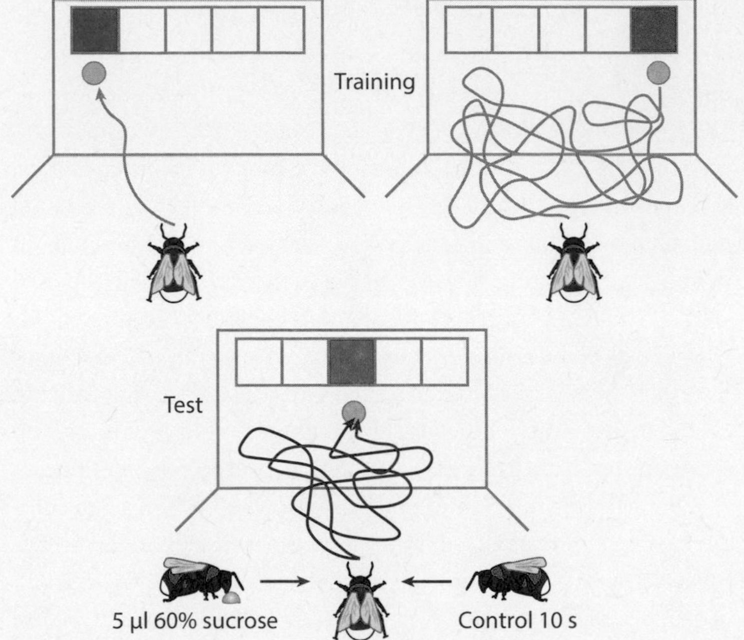


**Figure 11.3. An optimistic emotional bias in bumble bees.** Bumble bees familiarize themselves with a flight arena in which, on the back wall, the blue target on the left always contains a droplet of sugar water, whereas the green target on the right does not. Only one of the five positions on the choice wall is ever accessible. After training, bees will usually approach the blue target in a straight line when it is available (***top left***). Conversely, hesitation is apparent if the green target is on display (***top right***). When intermediate ("ambiguous") turquoise stimuli are offered (***bottom***), bees will hesitate for a time that lies between the flight times displayed for blue and green (black flight trajectory). If however, a tiny drop of sucrose solution (5 microliters) is offered as a surprise before bees enter the experiment, they judge the ambiguous option (turquoise) as more "promising" and will accept it more swiftly (red flight trajectory)—they judge the intermediate stimulus in a more optimistic manner than bees that had simply been held up on their way to the setup for 10 seconds.

be mood-altering. For example, male fruit flies experience ejaculation as rewarding—but if they are deprived of mating opportunities, they begin to seek out alcohol, which is often found in nature in the form of fermented fruits. Bees have been found to return preferentially to flowers with low levels of caffeine or nicotine in their nectar. Many plants naturally contain these substances in their foliage, because their bitter taste deters herbivores. However, these substances are sometimes found to leak into flower nectar at low concentrations as well, and this does not appear to be an accident. Their presence actually manipulates pollinator behavior to the plant's advantage, in that pollinators will return to such flowers even if their rewards become suboptimal. One might, of course, explain the effects of these psychoactive substances on the addiction-like behavior of pollinators by "simple" mechanisms—they might merely affect synaptic transmission in neural circuits that promote the learning of flower features, for example. But, given what we now know about the existence of emotional states in bees, it is equally plausible that bees are after these substances for the same reasons humans are—that they are actually mood-altering.

## Is Distinguishing Self-Generated from Other-Generated Sensory Stimuli at the Roots of the Evolution of Consciousness?

Some scholars now contend that elements of consciousness may have appeared near the origin of animal life, and were perhaps already found in the Cambrian (541–485 million years ago), close to the roots of the evolutionary tree branches that gave rise to modern arthropods and vertebrates. This is because some form of elementary self-recognition is required for most animals with self-generated, intentional movements. When you move, the picture you see changes. It might also change if you don't move, but then this is because something in the outside world has changed. This means that you need to take into account your own intentional movements to know and predict whether perceived change is a result of environmental events or of your own voluntary actions. If the image on your retina suddenly tilts by 45 degrees, you know that this is fine as long as it's the result of you deliberately cocking your head. But if you didn't move your head, you might be in the middle of an earthquake, and you'd better run.

Animals can tell the difference between these scenarios via what's known as an *efference copy*: an internal signal that communicates the consequences of an animal's own actions, so that it can distinguish sensory changes caused by its own movements from changes caused by external forces. Under normal conditions, animals expect the environment to appear to move in a predictable manner when they voluntarily turn their heads, or their whole bodies. This allows them to anticipate what will happen next as a result of their own actions or intentions. This would already have been necessary in the earliest animals that moved and actively explored their environment in search of food, actively running their touch and chemical sensors over their substrate, but also constantly on their guard for other creatures that might consume them. Even at this basic level of animal existence, a distinction of self from other (perhaps facilitated by a form of self-awareness) is essential. Once you evolve eyes, you have longer-distance sensors—but the basic challenge of distinguishing self- from other-generated sensory input remains.

Consider, for example, "looming" stimuli. If an object rapidly expands in your field of view, it means that it's coming at you (perhaps it's a predator, or in the human world, a speeding car). The natural response to such looming stimuli is to take evasive action. But if that was a hardwired response, a bee could never land on a flower: as the bee approaches its target, the flower grows larger and larger—a looming stimulus. But it is not experienced as threatening, because the bee "knows" that the expansion is not caused by something in the outside world, but by its own intentional actions. It may be theoretically possible to compute such differences between self- and other-generated visual motion cues without conscious awareness, but it has been suggested that computing this distinction may have actually given rise to the evolution of animal consciousness.

## The Self-Image of Bumble Bees

A popular test of animal self-awareness is mirror self-recognition: experimenters attach a color mark to the forehead of an animal, and then show it a mirror. Animals that touch their foreheads to try to remove the mark appear to recognize "myself" in the mirror, and to understand that their own appearance has suffered a blemish. This test is unlikely to work in bees, in part because their facial features do not vary sufficiently from one individual to another. However, evidence for a form of appreciation of the self comes from a study in which bumble bees had to take account of their own individual body dimensions to fly through small gaps.

We learned in chapter 10 that bumble bees of the same colony can differ widely in body size. Since bees don't grow once they have emerged from the pupa, their adult body size is constant for the remainder of their lives. As flower visitors, most bumble bees face daily demands to fly through dense vegetation without colliding with obstacles. This makes them ideal subjects to ask whether they are aware of their individual body size. In a new study, Sridhar Ravi and his team challenged bees with the task of flying through small openings of various sizes. Bees carefully scanned the openings before traversing them, apparently establishing how large the gap was. When the size of the gap was similar to a bee's individual wingspan or smaller, the bees angled their bodies or flew sidewise to traverse the gap, showing some form of knowledge of their own body dimensions (figure 11.4).

This is noteworthy since in other animals (including humans), knowing one's body dimensions is viewed as a core aspect of individual experience and self-awareness. An important next step will be to explore *how* bees learn about their own bodies to ensure that they can navigate safely through cluttered environments. It is possible that bees only learn this through trial and error after initiating their flight activities; but that might entail costly collisions with obstacles, during which their fragile wings could be damaged. It would be advantageous if bees could learn about their own body dimensions using their tactile sense during their early days in the hive, and later transfer this knowledge to the visual modality.

## Telling Self from Other Living Beings

The distinction of self from other is also important in another context—finding sex partners. All animals have some ability to recognize members of their own species and distinguish them from other species. Finding partners typically does not involve only recognition of members of the opposite sex of one's species: it is also important to identify individuals that are at least moderately dissimilar to oneself genetically, to avoid inbreeding. This, in turn, necessitates an assessment of the features that identify the self (often by olfactory cues), and a comparison of potential partners with these cues. Bees' ancestors would already have come with

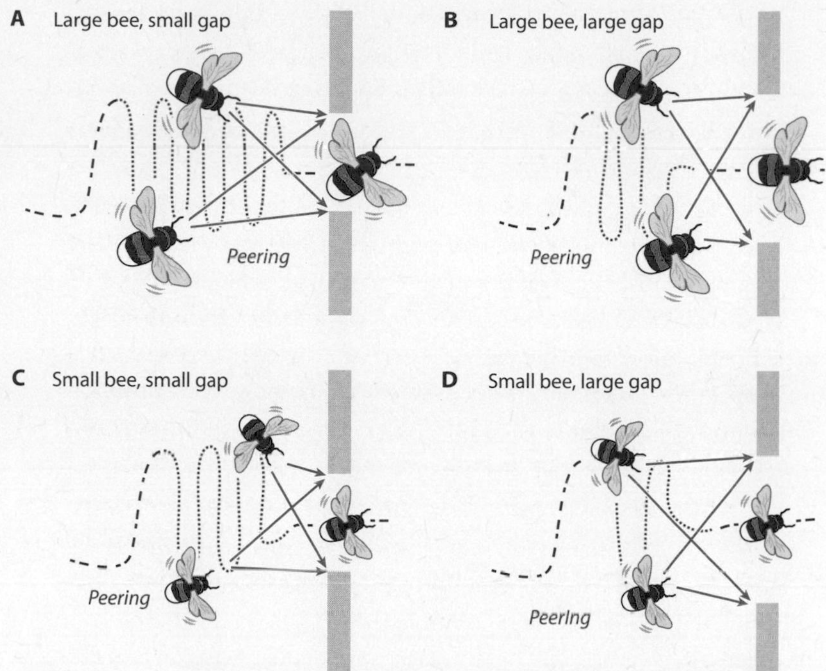


**Figure 11.4. Bees flying through small gaps show an appreciation of their own body dimensions.** During approach, bees examine the width of the gap by peering (red flight tracks), viewing the gap from several perspectives (blue arrows). When the gap is larger than the wingspan of the individual bee (***B.* and *D.***), bees fly straight through the center of the gap, but if the gaps are so small as to pose a risk of collision (***A.* and *C.***), bees will angle their bodies before navigating the gap, depending on their individual body size and wing span.

such abilities; in social bees, the olfactory recognition of colony members versus intruders from different colonies by guard bees is an extended form of distinguishing "self" from "other."

Such assessments involve more than just a "blind" comparison of one's own scent with that of another individual. There must be a basic form of recognition of what constitutes another individual.

Bees (and probably most animals) have a sense of the gestalt of conspecifics—where the basic anatomical features belong. While bumble bee males are shockingly indiscriminate in terms of the individuals they will attempt to mate with when an appropriate female is not in reach (workers, brothers, and queens of different species are all given a go), you will never see a male mounting a female back to front. Such basic recognition of body anatomy is essential whether another animal is a member of the same or of a different species. When a bear attacks a bee colony, for example, the bees alerted by alarm pheromone don't randomly attempt to sting anything in sight; their attacks are targeted at various body parts of the intruder—indicating a form of knowledge of where the body of their attacker begins and ends.

An inanimate object is something that ordinarily "moves" as a whole in a manner that is predictable from your own movement. A living being is something that moves as a whole, but which moves—and whose component parts move relative to each other—in ways not fully predictable from your own movements. Such a basic comprehension of what constitutes the parts of a single living being, whether it's a potential mate, a family member, or a predator, is probably widespread in the animal kingdom. At its root, it hinges on the earliest distinction of self from non-self—perceived changes in the environment that are due to one's own movements versus those generated by other animals.

## Offline Thinking in Bees?

Consciousness is a state of awareness that allows animals to not just live in the present, but also have access to the past and the future. Consciousness allows us to close our eyes and picture our childhood home, and it facilitates planning, prediction, and risk assessment, such as gauging whether it's safe to jump over a stream of a given width. There is evidence from psychological studies that these abilities are present in many animals in some form. Bees certainly appear to conjure up spatial memories of distant locations (including of their home; see chapter 6), and walking insects such as locusts (in their solitary phase) can visually estimate rung distances when walking on a ladder, and then plan their step width accordingly (even when the target rung is hidden from sight after the locust begins its step).

We have learned in the preceding paragraphs, and indeed across the entire book, that insects are far from non-sentient "automatons" without any form of internal representation of the world or any ability to foresee even the immediate future. The historical view—that in the absence of external stimulation or internal triggers such as hunger, the insect's mind is dark and its brain switched off—is no longer tenable. For example, we have heard about honey bees' abilities to retrieve spatial memories of feeding locations during the night, and to communicate about such locations outside of normal foraging times (chapter 6); we have seen a form of outcome awareness in bees' object manipulation skills (chapter 8). And we have tentative evidence that "brain waves"—the kinds of electrical oscillations, synchronized across brain regions, that form the neural hallmarks of consciousness in mammals—can also be found in insects (chapter 9).

## Imagining Shapes

One aspect of consciousness is an awareness of the sensory world around yourself. It may be hard to imagine that any animal with seeing eyes would not have some kind of representation of the outside world inside their brains, but this is surprisingly hard to prove. Bees can certainly respond to visual stimuli, and can learn to associate visual patterns (like those of flowers) with nectar rewards, but this does not necessarily imply that they have little virtual images of flowers floating around in their heads. A bee could store these complex visual patterns just by memorizing simple features, such as the orientation of edges, without actually storing full images in its memory—in other words, without an awareness of the actual patterns.

There is also evidence from humans that visual patterns can in principle be recognized without an awareness of the patterns. Patients with damage to their visual cortex sometimes lose all conscious visual experience—meaning they are experientially blind. When asked to locate a particular object, or discriminate between two visual patterns, they express no confidence that they can solve the task. But when asked to guess, they will still perform above chance level—a phenomenon called blindsight. Visual stimulus recognition is thus at least technically possible without awareness. Could it be that when the bee is hovering in front of the sought-after yellow flower, it somehow just "feels right," without the bee actually seeing a picture of the flower in its mind? A smartphone, after all, can recognize your face, but it does so without any form of awareness.

But new work on cross-modal object recognition indicates that something is going on inside the mind of bees that is wholly different from a machine—that bees *can* conjure up mental images of shapes. Cross-modal object recognition means that you, or an animal, can learn the features of an object in one sensory modality—such as vision—and then later recognize the same object in a different sensory modality—say, touch. For example, you can reach into a dark bag and identify objects by touch, even if previously you have only seen them—e.g., a ball, a pyramid-shaped object, or a die. The nerve signals that your eyes send to the brain are completely different from those sent by the touch sensors in your fingers, including in their temporal structure: you may need to run your fingers over an object for some seconds to ascertain its identity, whereas using vision you can recognize it at a glance. The reason you can nonetheless identify the correct object (even if you've never had to do so by touch alone) is that in your mind you can picture the key features of the object.

This ability to recognize shapes across sensory modalities is at the heart of the famous Molyneux problem, a seventeenth-century psychological conundrum. The Irish philosopher William Molyneux, whose wife was blind, wrote in 1688 to his English colleague John Locke the following question: "A Man, being born blind, and having a Globe and a Cube, nigh of the same bignes, Committed into his Hands, and being taught or Told, which is Called the Globe, and which the Cube, so as easily to distinguish them by his Touch or Feeling; Then both being taken from Him, and Laid on a Table, Let us Suppose his Sight Restored to Him; Whether he Could, by his Sight, and before he touch them, know which is the Globe and which the Cube?"

The question of whether previously blind humans could spontaneously solve this task when their sight was restored continues to puzzle psychologists to this day, in part because it is hard to answer in human subjects. Accordingly, we asked if bumble bees could identify shapes in the dark that they had previously only seen, or recognize shapes by vision that they had only felt with their tactile sense in the dark. We used the two shapes that Molyneux suggested, spheres and cubes (figure 11.5). One group of bees learned that spheres were associated with sugary rewards, and another that cubes were rewarding. These groups were further divided into bees that only ever experienced the objects in the dark, where they could touch but of course not see them, or in light, where they could see the objects through a plexiglass lid without being able to touch them (figure 11.5).

The results were striking: most bees had no difficulty spontaneously recognizing objects in a sensory modality in which they had never experienced them before. This indicates that the bees might indeed have a mental representation of the shape or the features of the object, rather than the patterns being recognized by simple feature detectors in their visual system (figure 11.5). Thus our bumble bees almost, but not quite, solved Molyneux's problem. Our bees were not "born blind," nor did they spend their entire time up to the beginning of our experiments in darkness. This means that they could have formed the links between how shapes feel and how they look by inspecting other objects using both their visual and tactile senses. However, given that bees' nests are naturally dark, it would be easy to test whether bees raised in wholly dark conditions and having learned to experience certain objects as rewarding or unrewarding, would spontaneously recognize those shapes the first time they see daylight.

Either way, these experiments demonstrate that bumble bees possess an ability to integrate sensory information that suggests they make modality-independent internal representations of objects. Perhaps, similar to humans and other large-brained animals, insects integrate information from multiple senses into a complete, globally accessible perception of the world around them.

**Figure 11.5. Can bees "picture" shapes in their mind?** Bumble bees in a cross-modal object recognition task: Bees learned to associate reward to one shape (here: a sphere) in a "look but don't touch" situation in the light, where a plexiglass lid prevented them from feeling the shape. Bees subsequently had to recognize the same shape in darkness, when they could feel but not see the shapes. The same experiment was also repeated the other way around, in which bees were first exposed to the objects in darkness, and subsequently had to find the right object shape in the "look but don't touch" situation.

## Do Bees Know What They Know?

Bees even appear to have metacognition—the ability to know what they know. This can be tested in animals by confronting them with a difficult visual discrimination task (for example, two very similar colors or patterns), and giving subjects not just the option to choose the correct (rewarding) or incorrect (unrewarding) target, but also a third response possibility: to opt out of participating in the task at all. Cwyn Solvi, working with the Australia-based British insect scientist Andrew Barron, discovered that indeed, as tasks became more difficult, honey bees increasingly chose this third option, as if they were aware of their own uncertainty. Such metacognition has been viewed as a hallmark of consciousness in apes and in dolphins, and if the opt-out behavior in mammals is taken as evidence of a self-assessment of uncertainty, then by the same criteria bees qualify too.

There has never been a formal proof of consciousness in any animal, and in this book I have not supplied a formal proof for bees, either. Critical readers might counter that every single psychological phenomenon, every intelligent behavior, described in this book could somehow be replicated by a computing algorithm or a robot, and therefore could in theory be accomplished without any form of conscious awareness. They would be right. You could design a robotic system for planning honeycomb construction, you could build robots that behave as if they experience pain when damaged, you could of course mimic the counting abilities of bees quite easily in silico. And the list goes on. But, first of all, if you wanted to build an automaton that could do *everything* I've described in this book—the dozens of "innate" as well as learned and innovated behaviors—you would have to equip your robot with a *very* long list of detailed instructions, and your machine would still be able to cope only with what you have programmed it to cope with. It would be helpless with any novel challenge for which you have not written any code.

The list of discoveries of intelligent behaviors displayed by bees is unlikely to end today. As Karl von Frisch highlighted in 1950: "The bee's life is like a magic well: the more you draw from it, the more it fills with water." He was right, and yet he would no doubt be astonished if we could travel back in time to update him on the breakthroughs that have been made in understanding the mental abilities of bees in the decades since his death in 1982. And the list of such abilities grows ever longer. So we have to ask ourselves if the historical notion of bees' nervous systems as a collection of cleverly tailored fixed circuits—perhaps one for each type of behavior—is really the "simpler" explanation. Perhaps a consciousness-based general intelligence system is not only more flexible in problem solving but also less computationally costly, and requires fewer nerve cells.

On balance, the evidence for at least a simple form of consciousness in bees is mounting. If we apply the same behavioral and cognitive criteria as we do to much larger-brained vertebrates, then bees qualify as conscious agents with no less certainty than dogs or cats. But the consciousness of bees might well be very different from that of humans. It is not (just) a question of the nature and richness of their sensory perception (where the differences among animal species are fairly obvious), or of whether some animals possess "more" or "less" consciousness. Different animals might differ profoundly in their perception of time and its use for the organization of memory and planning, in the appreciation of the self, and in many other dimensions. And this budding discipline of "comparative consciousness" opens up a fascinating perspective on bees' minds that we are only beginning to explore.

Perhaps bees have unique emotional states that accompany the excitement of the swarming process, or the thrill of discovering and successfully handling a particularly nectar-rich flower type, and the associated tapestry of multisensory experiences interacting with a rich library of multisensory memories—strange colors, odors, electrical signals from flowers, under the polarized light of the sky, as they zoom through meadows full of promising rewards but also potential threats. Further work is needed to explore these potential mind states, and their corresponding physiological states and behavioral expressions.

Animal brains construct information by establishing contingencies between sensory signals and their own actions, forming an internal model of the environment and of the self moving in it. This is what Charles Bonnet (see this chapter's epigraph), over 250 years ago, intuited when he commented about the "perfect resemblance of their outer and inner being" of the honey bee. From the very start, early in evolution, nervous systems were inseparable from movable bodies with sensors, and developed in order to integrate perception and action. The challenges of survival and self-replication (reproduction) that a moving organism faces are most efficiently met when brain and body are intimately connected, enabling the organism to know itself as distinct from non-self and to predict at least the immediate future, in part from knowing its own intentions. In this view, an elementary form of consciousness may have come into being at the roots, not at the endpoint, of animal life.

# 12

# Afterword

## What Our Knowledge of Bees' Minds Means for Their Conservation

In recent years, the global plight of bees has received much media attention. I am sometimes asked why bees can't cope with global change if they're so smart. Answer: they already do, on a grand scale.

Take a look at any arable flat landscape anywhere on the globe (you can inspect satellite images or, if you're privileged enough to take a plane occasionally, look out the window). The typical picture is that 99 percent of the landscape is dedicated to industrialized agriculture to feed seven billion people, and pasture for livestock to feed the non-vegetarians among us. The fragments of semi-natural landscape that remain are often pathetically tiny, and there are long distances between them. And the view from the plane does not even allow you to see the layers of insecticides and herbicides that cover the landscape, or the parasites and diseases that afflict bees as a result of indiscriminate shipping of pollinators around the globe. It is a testimony to animals' resilience and flexibility that so many species somehow still cling on. But there are limits to natural animals' adaptability—man-made change happens too rapidly for evolutionary change in most animals to keep up, and there are limits to what intelligence can do to compensate for extreme changes in your environment. Imagine humanity was faced with the challenge of losing over 90 percent of our living space over a few generations. Yes, some of us might survive, but not necessarily because of superior intelligence—more likely because of guns and money.

As for bees, the resilience of many species is already impressive. Shipped to distant continents for agricultural pollination, non-native bees swiftly learn to exploit local flowers, and often become invasive precisely because of their flexibility. They can on occasion also turn to completely non-natural food sources—for example, obtaining sugar boosts from discarded soft-drink cans. Maurice Maeterlinck reported in his century-old book that the honey bees of Barbados had given up flower visitation altogether and fed instead from the sugar harvested from the island's extensive sugar cane refineries.

Because of the lack of available natural nesting sites, honey bees nest in chimneys, and bumble bees in birds' nest boxes. Some solitary bees have been found building nests entirely from agricultural plastic waste (see figure 12.1), while others nest in polystyrene. But there are limits to the extent to which even the most intelligent animals can cope with extreme changes in their environment, when they occur too rapidly, and if competition between animals becomes extreme because of massive loss of habitat and suitable resources. Governments need to act swiftly to ban pesticides that have been proven to have detrimental effects on many beneficial insects, including bees, and must promote more re-naturalization of farmland.

However, in addition to governmental measures, a wonderful aspect of bee conservation is that every citizen with access to even small green spaces can contribute. Green lawns may be useful for

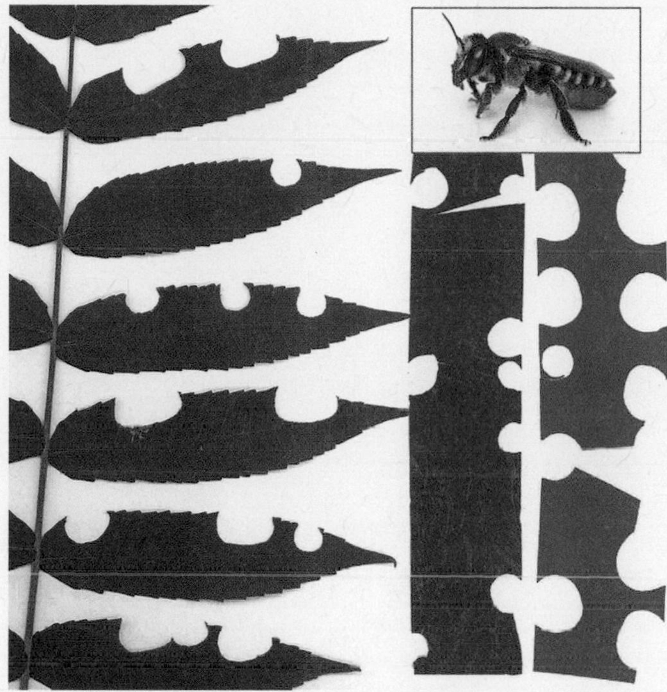

**Figure 12.1. Leafcutter bees of the genus *Megachile* sometimes use cuttings of plastic as nesting material.** Such bees normally use circular excisions of plant leaves (see left) as nesting material, but have been observed to utilize plastic flagging as a substitute. While this supports the extraordinary behavioral flexibility of bees, it is actually detrimental to the survival of their larvae.

certain sports, but they do not belong in gardens: they look dull, are labor-intensive, and are an ecological disaster. And many flowers have been bred to be large and showy to please human taste, but are entirely useless for pollinators. Instead, sow wildflowers, and let them grow among natural grasses. It is important to be selective in your flower choices; fortunately, this does not require laborious research. Many gardening centers now have pollinator-friendly seed mixes, and many such flowers "grow themselves"—all you have to do is sit back with a cool drink, watch the flowers grow happily in your garden, and discover the many native bees that may visit them. If you don't have a garden, even flower boxes on a balcony can help (especially if you consider them multiplied by the number of balconies in your town). Providing nesting opportunities for solitary bees is also a good idea—"bee hotels" are commercially available, but you can also manufacture your own, with expertise gleaned from just a bit of web searching.

Bear in mind that honey bee keeping is a lovely and informative hobby, but it is not a contribution to nature conservation. The western honey bee, to the extent that it is kept in hives, is a domesticated animal that, despite media reports to the contrary, is not under threat. A hive of 40,000 bees will deplete floral resources that could otherwise feed 40,000 wild solitary pollinators, many of which *are* at risk of extinction. (This is not even taking into account the fact that honey bees harvest more nectar and pollen than most other species of bees because of their habit of storing large quantities of food.) So: as with everything, moderation is important; a few hives kept by a few people is fine. However, the efforts of celebrities to "help pollinators" by setting up large numbers of honey bee colonies on their estates, or of many cities to encourage the placement of hundreds of hives on buildings' roofs, while well intentioned, are detrimental to many native pollinators.

We are often told that we need to conserve bees because we need them to pollinate our crops. That is undoubtedly the case, but I hold that our assessment of the possibility of pain, emotional states, and consciousness in bees also holds important implications for their conservation. We are currently as confident of these phenomena in bees as we are in the charismatic mammals that are the classic mascots of conservation campaigns—which in many cases enjoy their special status in part because we worry, for good reasons, that they might be conscious of, and suffer from, the destruction of their habitat and the disruption of their families and societies.

The assessment of pain and emotional states in animals is a moral obligation in deciding on strategies for animal welfare, for domestic animals and those hunted in the wild, as well as for animals used in laboratory experiments. However, current legislation in most countries places no limits on the treatment of invertebrates, on the assumption that they do not experience pain or emotions. This means that lobsters can be cooked alive in restaurants, and that tethered insects can be subject to invasive neurobiological procedures in the laboratory without anesthesia or analgesia. Indeed, there is no requirement to apply for a permit to ensure that the research is conducted with due consideration of the numbers of animals used, the severity of the experimental treatment, or the possible benefits to science or humanity.

Given what we have learned above about the strong likelihood that bees have more than basic nociception, and at least an elementary emotional life, this is clearly a deplorable state of affairs. It is thus imperative to conduct studies that provide more-comprehensive evidence of whether or not insects can experience the kind of suffering that would necessitate amendment of the animal welfare laws to include them and other invertebrates. In the meantime, we should err on the side of caution and treat bees with the same respect as other animals in which we accept the possibility of subjective experience.

And if you are still not swayed that bees have such experiences—and that this is an important argument for their conservation—then remember that we still need them to pollinate our crops, that the candles made from their wax have illuminated scholarly efforts through the ages, and that the harvesting of their honey might well have provided the necessary carbohydrates for our ancestors' costly brain enlargement, and might thus have fueled the evolution of the human mind. We owe the bees. Act accordingly.

# ACKNOWLEDGMENTS

The writing of this book was made possible by a generous fellowship from the Wissenschaftskolleg / Institute of Advanced Study in Berlin, in the 2017/18 academic year. I am exceedingly grateful to my editor Alison Kalett at Princeton University Press for her guidance and wisdom, and to three anonymous referees for their comprehensive and helpful feedback on the book manuscript. I am also indebted to Raghavendra Gadagkar and Janna Klein for each reading a chapter and providing comments. Several other people have greatly helped in turning the manuscript into a final product, most notably Annie Gottlieb, Amelia Kowalewska, Chris Lapinski, Hallie Schaeffer, Julie Shawvan, and Jenny Wolkowicki.

I thank my mentors Randolf Menzel and James Thomson for introducing me to the magic worlds of honey bees and bumble bees, respectively. I am grateful for the many scientists who joined my team from all over the world, shunning the fashionable science topics that get you easy careers in academia, and who instead were solely motivated by the curiosity of what's on the mind of an insect. Those who contributed significantly to the findings reported in this book are: Sylvain Alem, Sarah Arnold, Aurore Avarguès-Weber, Joanna Brebner, Erika Dawson, Anna Dornhaus, Adrian Dyer, Vince Gallo, Marie Guiraud, Thomas Ings, Ellouise Leadbeater, Li Li, Mathieu Lihoreau, Olli Loukola, HaDi MaBouDi, James Makinson, Helene Muller, Vivek Nityananda, Fei Peng, Nigel Raine, Mark Roper, Nehal Saleh, Cwyn Solvi, Johannes Spaethe, Ralph Stelzer, Vera Vasas, Mu-Yun Wang, Joseph Woodgate, and Xing-Fu Zhu.

The intellectual exchange with scientists from other teams has also been important for the ideas presented in this book—most importantly these are Adriana Briscoe, Thomas Collett, Karl Geiger, Martin Giurfa, Beverley Glover, Andreas Gumbert, Jan Kunze, Miriam Lehrer, Martin Lindauer, Jeremy Niven, Avi Shmida, Peter Skorupski, Nick Waser, Heather Whitney, and Neil Williams.

I also thank the following individuals for the generation of, and/or permission to use, or modify, their image materials: Sylvain Alem, Johanna Brebner, Brigitte Bujok, Jeremy Early, Vince Gallo, Andy Giger, Beverley Glover, Helga Heilmann, Scott Hodges, Thomas Ings, Steve Johnson, Marco Kleinhenz, Li Li, Beau Lotto, Iida Loukola, Klaus Lunau, HaDi MaBouDi, Rob Raguso, Stuart Roberts (IBRA) on behalf of Leslie Goodman, Rotraut Sachs, Florian Schiestl, Klaus Schmitt, Cwyn Solvi, Johannes Spaethe, Jürgen Tautz, Rüdiger Wehner, Joseph Wilson, and Joseph Woodgate. In the early stages of manuscript writing, I saw a magnificent photo of an orchid bee on social media, and I knew at once that this would have to be on the cover of the book. It was by the extraordinary photographer Andreas Kay (1963–2019), who tragically died soon after. He had captured insects with a unique artistic handwriting, highlighting that if we only zoom in sufficiently on these tiny creatures, we find a fascinating and alien world just on our doorstep. He thus achieved with pictures what I am hoping to accomplish in writing.

# NOTES AND BIBLIOGRAPHY


**Chapter 1**

1: Quote by Maeterlinck: from Maeterlinck, M. 1901. *La vie des abeilles.* Paris: Editions Frasquelle; trans. Alfred Sutro. 1903. New York: Dodd, Mead.

3: Estimate of cell number in the human brain: see Herculano-Houzel, S. 2009. "The human brain in numbers: a linearly scaled-up primate brain." *Frontiers in Human Neuroscience* 3 (31). DOI: 10.3389/neuro.09.031.2009.

3: Estimates of cell numbers in the brains of bees and other Hymenopteran insects: Witthöft, W. 1967. "Absolute Anzahl und Verteilung der Zellen im Hirn der Honigbiene." *Zeitschrift für Morphologie der Tiere* 61: 160–84; Godfrey, R. K., Swartzlander, M., Gronenberg, W. 2021. "Allometric analysis of brain cell number in Hymenoptera suggests ant brains diverge from general trends." *Proceedings of the Royal Society B-Biological Sciences* 288 (1947). DOI: 10.1098/rspb.2021.0199.

3: The idea of brains as prediction machines is further developed in: Heisenberg, M. 2015. "Outcome learning, outcome expectations, and intentionality in Drosophila." *Learning & Memory* 22 (6). DOI: 10.1101/lm.037481.114; and Menzel, R. 2019. "Search strategies for intentionality in the honeybee brain." In *Oxford Handbook of Invertebrate Neurobiology*, ed. J. H. Byrne, 663–84. DOI: 10.1093/oxfordhb/9780190456757.013.27. Oxford, UK: Oxford Handbooks Online.

3: "What it's like to be a bee" is a play on the title of an influential essay on the difficulty of understanding other animals' minds: Nagel, T. 1974. "What is it like to be a bat?" *The Philosophical Review* 83: 435–50. DOI: 10.2307.2183914.

4: The idea of taking the view from the cockpit of an insect was, to my knowledge, first suggested in Borst, A., Egelhaaf, M. 1992. "Im Cockpit der Fliege." *MPG-Spiegel* 3: 14–17. For a more up-to-date essay see Borst, A. 2009. "Drosophila's view on insect vision." *Current Biology* 19: R36–47. DOI: 10.1016/j.cub.2008.11.001.

4: For a comprehensive treatment of the optics of insect eyes: see Land, M. F., Nilsson, D.-E. 2002. *Animal Eyes*. Oxford: Oxford University Press; and, for a short overview: Land, M., Chittka, L. 2013. "Vision." In *The Insects: Structure and Function*, 5th edition, eds. S. J. Simpson and A. E. Douglas. Cambridge, UK: Cambridge University Press, 708–37; on the remarkable speed of insect visual information processing, see Niven, J. E., Anderson, J. C., Laughlin, S. B. 2007. "Fly photoreceptors demonstrate energy-information trade-offs in neural coding." *PLOS Biology* 5: e116. DOI: 10.1371/journal.pbio.0050116.

4: On the challenge of being a flower forager, a very good overview is provided by Heinrich, B. 1979. *Bumblebee Economics*. Cambridge, MA: Harvard University Press.

4: On how the evolutionary past, as well as individually acquired information, forms the contents of a mind: see Lorenz, K. 1978. *Behind the Mirror: A Search for a Natural History of Human Knowledge*. New York: Harcourt Brace Jovanovich.

6: On flowers having different meanings for bees and humans: Chittka, L., Walker, J. 2006. "Do bees like Van Gogh's *Sunflowers*?" *Optics and Laser Technology* 38: 323–28. DOI: 10.1016/j.optlastec.2005.06.020.

6: Very high loss rates on the very first flights of bees were observed in: Stelzer, R. J., Raine, N. E., Schmitt, K. D., Chittka, L. 2010. "Effects of aposematic coloration on predation risk in bumblebees? A comparison between differently coloured populations, with consideration of the ultraviolet." *Journal of Zoology* 282 (2): 75–83. DOI: 10.1111/j.1469-7998.2010.00709.x.

8: On how humans fail to orient adequately in the wilderness: see Bond, M. 2020. "People who get lost in the wild follow strangely predictable paths." *New Scientist* (3271), February 29, 2020.

10: Flowers as puzzle boxes, and how bees learn to open and manipulate them: Laverty, T. M., Plowright, R. C. 1988. "Flower handling by bumblebees: a comparison of specialists and generalists." *Animal Behaviour* 36: 733–40. DOI: 10.1016/S0003-3472(88)80156-8.

13: A honey bee's 15 pheromone glands: Free, J. B. 1987. *Pheromones of Social Bees*. Ithaca, NY: Comstock; Blum, M. S. 1992. "Honey bee pheromones," in *The Hive and the Honey Bee*, revised edition (Hamilton, Illinois: Dadant and Sons), 385–89.

13: The most comprehensive exploration of the honey bee dance language is: von Frisch, K. 1967. *The Dance Language and Orientation of Bees*. Cambridge, MA: Harvard University Press.

14: Philosophers' take on what it's like to be another animal: see Nagel.

15: On emotions in bees and other invertebrates: Perry, C. J., Baciadonna, J. 2017. "Studying emotion in invertebrates: what has been done, what can be measured and what they can provide." *Journal of Experimental Biology* 220 (21): 3856–3868. DOI: 10.1242/jeb.151308.

16: Males (in bees) are only good for sex; an exception is found in some bumble bees where males appear to participate in warming the brood: Cameron, S.A. 1985. "Brood care by male bumblebees." *Proceedings of the National Academy of Sciences of the USA* 82 (19): 6371–73. DOI: 10.1073/pnas.82.19.6371.

17: Further reading on ethical considerations: Mikhalevich, I., Powell, R. 2020. "Minds without spines: evolutionarily inclusive animal ethics." *Animal Sentience* 329: 1–25. DOI: 10.51291/2377-7478.1527.

18: On the long shared history of humans and bees, see: Stanford, C. B., Gambaneza, C., Nkurunungi, J. B., Goldsmith, M. L. 2000. "Chimpanzees in Bwindi-Impenetrable

National Park, Uganda, use different tools to obtain different types of honey." *Primates* 41 (3): 337–41. DOI: 10.1007/bf02557602; Marlowe, F. W., Berbesque, J. C., Wood, B., Crittenden, A., Porter, C., Mabulla, A. 2014. "Honey, Hadza, hunter-gatherers, and human evolution." *Journal of Human Evolution* 71: 119–28. DOI: 10.1016/j.jhevol.2014.03.006; and for popular scientific overviews: Hanson, T. 2018. *Buzz*. New York: Basic Books; and Preston, C. 2006. *Bee*. London: Reaktion Books.

19: Details about the remarkable life and science of Charles Turner can be found in: Abramson, C. I. 2009. "A study in inspiration: Charles Henry Turner (1867–1923) and the investigation of insect behavior." *Annual Review of Entomology* 54: 343–59. DOI: 10.1146/annurev.ento.54.110807.090502; Wehner, R. 2016. "Early ant trajectories: spatial behaviour before behaviourism." *Journal of Comparative Physiology A–Sensory, Neural, and Behavioral Physiology* 202 (4): 247–66. DOI: 10.1007/s00359-015-1060-1; Lee, D. N. 2020. "Diversity and inclusion activisms in animal behaviour and the ABS: a historical view from the USA." *Animal Behaviour* 164: 273–80. DOI: 10.1016/j.anbehav.2020.03.019; Galpayage Dona, H. S., Chittka, L. 2020. "Charles H. Turner, pioneer in animal cognition." *Science* 370 (6516): 530–31. DOI: 10.1126/science.abd8754.

**Chapter 2**

20: Introductory quote by Lord Rayleigh: Lord Rayleigh (Strutt, J. W.). 1874. "Insects and the colours of flowers." *Nature* 11: 6. DOI: 10.1038/011006a0.

21: John Lubbock's experiments on ants' UV sensitivity, and bee color training: Lubbock, J. 1882. *Ants, Bees and Wasps: A Record of Observations on the Habits of Social Hymenoptera*. London: Kegan Paul, Trench, Trubner, and Co., 442pp.

21: In addition to Lubbock's experiments on color learning in bees, such tests were also performed by Charles Turner. Turner, C. H. 1910. "Experiments on color-vision of the honey bee." *Biological Bulletin* 19: 257–79. Like Lubbock, Turner did not control for intensity of stimuli, but he pointed out that such a control would have been desirable. For further historical context, see: Giurfa, M., Sanchez, M.G.D. 2020. "Black lives matter: revisiting Charles Henry Turner's experiments on honey bee color vision." *Current Biology* 30 (20): R1235–39. DOI: 10.1016/j.cub.2020.08.075.

22: First book on color vision in animals: von Hess, C. 1912. *Vergleichende Physiologie des Gesichtssinnes*. Jena: G. Fischer.

22: Von Hess' experiment on bee color vision: von Hess, C. 1913. "Experimentelle Untersuchungen über den angeblichen Farbensinn der Bienen." *Zoologische Jahrbücher* 34: 81–106.

24: Von Frisch's seminal paper on bee color vision: von Frisch, K. 1914. "Der Farbensinn und Formensinn der Biene." *Zoologische Jahrbücher (Physiologie)* 37: 1–238; the controversy between von Frisch and von Hess is described in detail in Kreutzer, U. 2010. *Karl von Frisch—eine Biografie*. München: August Dreesbach Verlag.

25: Von Frisch's autobiography: von Frisch, K. 1973. *Erinnerungen eines*

*Biologen*. Berlin, Heidelberg, New York: Springer.

26: Discovery of UV sensitivity in bees: Kühn, A. 1923. "Versuche über das Unterscheidungsvermögen der Bienen und Fische für Spektrallichter." *Nachrichten von der Gesellschaft der Wissenschaften zu Göttingen, Mathematisch-Physikalische Klasse*: 66–71.

26: Discovery of UV reflectance in flowers: Lutz, F. E. 1924. "Apparently non-selective characters and combinations of characters including a study of ultraviolet in relation to the flower-visiting habits of insects." *Annals of the New York Academy of Sciences* 29: 181–283.

26: Karl von Frisch and the Nazis; for details see the autobiography, above, plus the Kreutzer biography cited above; in addition, see Munz, T. 2016. *The Dancing Bees: Karl von Frisch and the Discovery of the Honeybee Language*. Chicago: University of Chicago Press.

26: "second degree mongrel" quote from the Nazi Bavarian State Ministry: see Kreutzer biography, above.

26: "bigoted opposition to antisemitism": see von Frisch autobiography, above.

26: Karl von Frisch's book *You and Life*: von Frisch, K. 1936. *Du und das Leben: Eine moderne Biologie*. Berlin: Im Deutschen Verlag.

28: Karl Daumer's discoveries on bee color vision: Daumer, K. 1956. "Reizmetrische Untersuchung des Farbensehens der Bienen." *Zeitschrift für Vergleichende Physiologie* 38: 413–78. DOI: 10.1007/BF00340456.

28: On how color vision is fundamentally different from, e.g., odor and sound perception: Chittka, L., Brockmann, A. 2005. "Perception space, the final frontier." *PLOS Biology* 3: 564–68. DOI: 10.1371/journal.pbio.0030137.

31: Hansjochem Autrum's intracellular recordings of bee photoreceptors: Autrum, H. J., Zwehl, V.v. 1964. "Die spektrale Empfindlichkeit einzelner Sehzellen des Bienenauges." *Zeitschrift für Vergleichende Physiologie* 48: 357–84. DOI: 10.1007/BF00299270.

32: Randolf Menzel's work on the speed of color learning in bees: Menzel, R. 1985. "Learning in honey bees in an ecological and behavioral context." In *Experimental Behavioral Ecology* (eds. Hölldobler, B., Lindauer, M.), 55–74. Stuttgart: Gustav Fischer Verlag.

32: On learning speed not being a useful measure of intelligence: Pearce, J. M. 2008. *Animal Learning and Cognition*. 3rd edition Hove, UK, and New York: Psychology Press.

33: Did bee color vision evolve in response to flower colors? See: Chittka, L., Menzel, R. 1992. "The evolutionary adaptation of flower colors and the insect pollinators' color vision systems." *Journal of Comparative Physiology A* 171: 171–81. DOI: 10.1007/BF00188925; Chittka, L. 1996. "Optimal sets of colour receptors and opponent processes for coding of natural objects in insect vision." *Journal of Theoretical Biology* 181: 179–96. DOI: 10.1006/jtbi.1996.0124.

37: Phylogenetic analyses of insect color vision: Chittka, L. 1996. "Does bee colour vision predate the evolution of flower colour?" *Naturwissenschaften* 83: 136–38. DOI: 10.1007/BF01142181; Briscoe, A., Chittka, L. 2001. "The evolution of colour vision in insects." *Annual Review of Entomology* 46: 471–510. DOI: 10.1146

/annurev.ento.46.1.471; van der Kooi, C. J., Stavenga, D. G., Arikawa, K., Belušič, G., Kelber, A. 2021. "Evolution of insect color vision: from spectral sensitivity to visual ecology." *Annual Review of Entomology* 66 (1): 435–61. DOI: 10.1146/annurev-ento-061720-071644.

**Chapter 3**

39: Introductory quote by John Lubbock: Lubbock, J. 1888. "Problematical organs of sense." *Popular Science Monthly* 34: 101–7. The quote continues: "To place stuffed birds and beasts in glass cages, to arrange insects in cabinets . . . is merely the drudgery and preliminary of study; to watch their habits, to understand their relations to one another, to study their instincts and intelligence, to ascertain their adaptations and their relations to the forces of nature, to realise what the world appears to them—these constitute . . . the true interest of natural history."

40: John Lubbock's parliamentary duties, entomology, ants' chemical language, UV sensitivity, and bee color training: Lubbock, J. 1882. *Ants, Bees and Wasps: A Record of Observations on the Habits of Social Hymenoptera.* London: Kegan Paul, Trench, Trubner, and Co, 442pp.

40: On the interactions between John Lubbock, Alexander Graham Bell, and Charles Darwin: Keynes, R. 2009. "'I thought I'd try the telephone'—Darwin, his disciple, insects and earthworms." *Journal of the Linnean Society* Special Issue 9: 79–96.

41: Speed of insect vision: Srinivasan, M., Lehrer, M. 1985. "Temporal resolution of colour vision in the honeybee." *Journal of Comparative Physiology A* 157: 579–86. DOI: 10.1007/BF01351352; Niven, J. E., Laughlin, S. B. 2008. "Energy limitation as a selective pressure on the evolution of sensory systems." *Journal of Experimental Biology* 211 (11): 1792–1804. DOI: 10.1242/jeb.017574; Skorupski, P., Chittka, L. 2010. "Differences in photoreceptor processing speed for chromatic and achromatic vision in the bumblebee, *Bombus terrestris*." *Journal of Neuroscience* 30 (11): 3896–903. DOI: 10.1523/jneurosci.5700–09.2010.

41: "Sense organs can be in strange places in insects": Yager, D. D. 1999. "Structure, development, and evolution of insect auditory systems." *Microscopy Research and Technique* 47 (6): 380–400. DOI: 10.1002/(sici)1097-0029(19991215)47:6<380::aid-jemt3>3.0.co;2-p; Arikawa, K., Eguchi, E., Yoshida, A., Aoki, K. 1980. "Multiple extraocular photoreceptive areas on genitalia of butterfly *Papilio xuthus*." *Nature* 288: 700–702. DOI: 10.1038/288700a0.

42: Biographical details of Martin Lindauer: Seeley, T. D., Kühnholz, S., Seeley, R. H. 2002. "An early chapter in behavioral physiology and sociobiology: the science of Martin Lindauer." *Journal of Comparative Physiology A* 188: 439–53. DOI: 10.1007/s00359-002-0318-6.

44: Ernst Wolf's work on the bees' sun compass: Wolf, E. 1927. "Über das Heimkehrvermögen der Bienen II." *Zeitschrift für Vergleichende Physiologie* 6: 221–54.

46: Martin Lindauer's work with Karl von Frisch on the sun compass: Lindauer, M. 1985. "Karl Ritter von Frisch, 1886–1982." In *Die großen Deutschen unserer Epoche*, ed. Lothar Gall, 453–65. Berlin: Propyläen Verlag.

47: Martin Lindauer's work on bees predicting the sun's course is described in detail in von Frisch, *The Dance Language and Orientation of Bees*.

49: Von Frisch's work (jointly with Martin Lindauer) on the discovery of bee polarization vision was performed on bees' communication dances in the hive (the dance language is described in chapter 5 of this book). References for the original work are in von Frisch, *The Dance Language*.

50: Rüdiger Wehner's work on the mechanisms of polarization vision in bees: Wehner, R., Bernard, G. D., Geiger, E. 1975. "Twisted and non-twisted rhabdoms and their significance for polarization detection in the bee." *Journal of Comparative Physiology* 104: 225–45. DOI: 10.1007/BF01379050; Rossel, S., Wehner, R. 1982. "The bee's map of the e-vector pattern in the sky." *Proceedings of the National Academy of Sciences of the USA* 79: 4451–55. DOI: 10.1073/pnas.79.14.4451; Wehner, R., Labhart, T. 2006. "Polarisation vision." In *Invertebrate Vision*, ed. E. J. Warrant, D.-E. Nilsson, 291–348. Cambridge, UK: Cambridge University Press.

52: Lindauer's work on bees' sensitivity to magnetic fields was performed on dancing bees inside the hive: Lindauer, M., Martin, H. 1968. "Die Schwereorientierung der Bienen unter dem Einfluß des Erdmagnetfeldes." *Zeitschrift für Vergleichende Physiologie* 60: 219–43. DOI: 10.1007/BF00298600.

52: Further evidence for sensitivity of insects to the Earth's magnetic field: Gould, J. L., Kirschvink, J. L., Deffeyes, K. S. 1978. "Bees have magnetic remanence." *Science* 201: 1026–28. DOI: 10.1126/science.201.4360.1026; Frier, H. J., Edwards, E., Smith, C., Neale, S., Collett, T. S. 1996. "Magnetic compass cues and visual pattern learning in honeybees." *The Journal of Experimental Biology* 199: 1353–61. DOI: 10.1242/jeb.199.6.1353; Gegear, R. J., Casselman, A., Waddell, S., Reppert, S. M. 2008. "Cryptochrome mediates light-dependent magnetosensitivity in *Drosophila*." *Nature* 454 (7207): 1014-18. DOI: 10.1038/nature07183; Dreyer, D., Frost, B., Mouritsen, H., Gunther, A., Green, K., Whitehouse, M., Johnsen, S., Heinze, S., Warrant, E. 2018. "The Earth's magnetic field and visual landmarks steer migratory flight behavior in the nocturnal Australian Bogong moth." *Current Biology* 28 (13): 2160–66.e5. DOI: 10.1016/j.cub.2018.05.030; Wajnberg, E., Acosta-Avalos, D., Alves, O. C., de Oliveira, J. F., Srygley, R. B., Esquivel, D.M.S. 2010. "Magnetoreception in eusocial insects: an update." *Journal of the Royal Society Interface* 7: S207–25. DOI: 10.1098/rsif.2009.0526.focus.

53: Experiments on bee orientation in darkness: Chittka, L., Williams, N. M., Rasmussen, H., Thomson, J. D. 1999. "Navigation without vision: bumblebee orientation in complete darkness." *Proceedings of the Royal Society of London B* 266: 45–50. DOI: 10.1098/rspb.1999.0602.

53: On how suboptimal working conditions can sometimes lead to surprising insights (and other suggestions for how scientists achieve extraordinary breakthroughs): see the following essay by Richard Hamming: Hamming, R. 1986. "You and your research." Transcript of the Bell Communications Research Colloquium Seminar, March 7, 1986. Morristown, NJ: Bell Communications Research.

53: Mechanisms of magnetic field sensitivity: Liang, C. H., Chuang, C. L., Jiang, J. A., Yang, E. C. 2016. "Magnetic sensing through the abdomen of the honey bee." *Scientific Reports* 6. DOI: 10.1038/srep23657.

54: An excellent overview of the many functions of the bees' antennae is in: Goodman, L. 2003. *Form and Function in the Honeybee*. Cardiff, UK: Westdale Press.

55: Numbers of and types of antennal receptors in bees: Esslen, J., Kaissling, K. E. 1976. "Zahl und Verteilung antennaler Sensillen bei der Honigbiene (*Apis mellifera* L.)." *Zoomorphologie* 83: 227–51. DOI: 10.1007/BF00993511.

56: Further reading on $CO_2$ sensitivity in honey bees and other insects: Seeley, T. D. 1974. "Atmospheric carbon-dioxide regulation in honeybee (*Apis mellifera*)." *Journal of Insect Physiology* 20: 2301–5. DOI: 10.1016/0022-1910(74)90052-3; Jones, W. 2013. "Olfactory carbon dioxide detection by insects and other animals." *Molecular Cell* 35 (2): 87–92. DOI: 10.1007/s10059-013-0035-8.

56: Number of volatiles in a single plant species' bouquet: Friberg, M., Schwind, C. Guimarães P. R., Jr., Raguso, R. A., Thompson, J. N. 2019. "Extreme diversification of floral volatiles within and among species of *Lithophragma* (Saxifragaceae)." *Proceedings of the National Academy of Sciences of the USA* 116 (10): 4406–15. DOI: 10.1073/pnas.1809007116.

57: Randolf Menzel on odor learning and alarm pheromone: Menzel, R. 1985. "Learning in honey bees in an ecological and behavioral context." In *Experimental Behavioral Ecology* (eds. Hölldobler, B., Lindauer, M.), 55–74. Stuttgart: Gustav Fischer Verlag.

57: Bees as sniffers: Kerk, W. C., Chua, L. S. 2016. "Sniffer bees as a good alternative for the current sniffing technology." *Biointerface Research in Applied Chemistry* 6 (4): 1391–1400.

58: Speed of odor perception: Szyszka, P., Gerkin, R. C., Galizia, C. G., Smith, B. H. 2014. "High-speed odor transduction and pulse tracking by insect olfactory receptor neurons. *Proceedings of the National Academy of Sciences of the USA* 111 (47):16925–30. DOI: 10.1073/pnas.1412051111.

58: Karl von Frisch on bees' sense of taste: von Frisch, K. 1934. "Über den Geschmackssinn der Bienen." *Zeitschrift für Vergleichende Physiologie* 21: 1–156.

59: Bees' preference for nectar laced with pesticides: Kessler, S. C., Tiedeken, E. J., Simcock, K. L., Derveau, S., Mitchell, J., Softley, S., Stout, J. C., Wright, G. A. 2015. "Bees prefer foods containing neonicotinoid pesticides." *Nature* 521 (7550): 74–76. DOI: 10.1038/nature14414.

59: Touch sensors on bees' antennae and their function in sensing floral texture: Kevan, P. G., Lane, M. A. 1985. "Flower petal microtexture is a tactile cue for bees." *Proceedings of the National Academy of Sciences of the USA* 82: 4750–52. DOI: 10.1073/pnas.82.14.4750; Whitney, H. M.,

Chittka, L., Bruce, T.J.A., Glover, B. J. 2009. "Conical epidermal cells allow bees to grip flowers and increase foraging efficiency. *Current Biology* 19 (11): 948–53. DOI: 10.1016/j.cub.2009.04.051.

60: Overview of hearing in insects: Robert, D., Gopfert, M. C. 2002. "Novel schemes for hearing and orientation in insects." *Current Opinion in Neurobiology* 12 (6): 715–20. DOI: 10.1016/s0959–4388(02)00378–1.

60: Hearing in honey bees: Dreller, C., Kirchner, W. H. 1993. "Hearing in honeybees: localization of the auditory sense organ." *Journal of Comparative Physiology A* 173: 275–79. DOI: 10.1007/BF00212691; Kirchner, W. H., Towne, W. F. 1994. "The sensory basis of the honeybee's dance language." *Scientific American* 270: 74–81; Towne, W. F., Kirchner, W. H. 1989. "Hearing in honey bees: detection of air-particle oscillations." *Science* 244: 686–88. DOI: 10.1126/science.244.4905.686.

60: Sensing vibrations of the honeycomb with the legs: Nieh, J. C., Tautz, J. 2000. "Behaviour-locked signal analysis reveals weak 200–300 Hz comb vibrations during the honeybee waggle dance." *The Journal of Experimental Biology* 203: 1573–79. DOI: 10.1242/jeb.203.10.1573.

61: On the electrical properties of hair and feathers: Exner, S. 1895. "Über die elektrischen Eigenschaften der Haare und Federn." *Pflügers Archiv* 61: 1–98.

61: Bees' electrical sensitivity and its possible role in bee communication: Eskov, E. K., Sapozhnikov, A. M. 1974. "Generation and perception of electric fields by *Apis mellifera*." *Zoologičeskij žurnal* 52: 800–802; Eskov, E. K., Sapozhnikov, A. M. 1976. "Mechanisms of generation and perception of electric fields by honey bees." *Biofizika* 21 (6): 1097–1102; Greggers, U., Koch, G., Schmidt, V., Durr, A., Floriou-Servou, A., Piepenbrock, D., Gopfert, M. C., Menzel, R. 2013. "Reception and learning of electric fields in bees." *Proceedings of the Royal Society B-Biological Sciences* 280 (1759): 8. DOI: 10.1098/rspb.2013.0528.000; detection of floral electric fields by bumble bees: Sutton, G. P., Clarke, D., Morley, E. L., Robert, D. 2016. "Mechanosensory hairs in bumblebees (*Bombus terrestris*) detect weak electric fields." *Proceedings of the National Academy of Sciences of the USA* 113 (26): 7261–65. DOI: 10.1073/pnas.1601624113; Clarke, D., Whitney, H., Sutton, G., Robert, D. 2013. "Detection and learning of floral electric fields by bumblebees." *Science* 340 (6128): 66–69. DOI: 10.1126/science.1230883.

**Chapter 4**

64: Quote from Huber, F. 1814. *Nouvelles observations sur les abeilles* (2nd edition); trans. C.P. Dadant, as *New Observations upon Bees*. 1926. Hamilton, IL: *American Bee Journal*.

65: On the instinctual nature of human language: Pinker, S. 1994, *The Language Instinct*. New York: William Morrow.

66: The number and diversity of instinctual behaviors in bees: Chittka, L., Niven, J. 2009. "Are bigger brains better?" *Current Biology* 19: R995–1008. DOI: 10.1016/j.cub.2009.08.023.

67: Fabre on pine processionary caterpillars: Fabre, J.-H. 1900. *Souvenirs Ento-*

*mologiques—VIIe série*. Paris: Charles Delagrave.

68–69: "Fragments on the psychology of insects" and "You are nothing": Fabre, J.-H. 1882. *Nouveaux Souvenirs Entomologiques—IIe série*. Paris: Charles Delagrave.

70: Thoracic ganglia fused together: Niven, J. E., Graham, C. M., Burrows, M. 2008. "Diversity and evolution of the insect ventral nerve cord." *Annual Review of Entomology* 53: 253–71. DOI: 10.1146/annurev.ento.52.110405.091322.

70: Fabre's observations on digger wasps: Fabre, J.-H. 1879. *Souvenirs Entomologiques—Ire série*. Paris, Charles Delagrave.

70: Daniel Dennett on the "mindless mechanicity" of insect behavior: Dennett, D. C. 1984. *Elbow Room: The Varieties of Free Will Worth Wanting.* Cambridge, MA: MIT Press.

73: Comb-making power: Darwin, C. 1859. *The Origin of Species*, chapter 7: "Instinct." London: John Murray.

74: Huber, Lullin, and Burnens's experiments on flexibility of bee comb construction: Huber, F. 1814. *Nouvelles observations sur les abeilles (seconde édition)*—trans. C.P. Dadant, as *New Observations upon Bees*. 1926. Hamilton, IL: *American Bee Journal*; for a more recent discussion of this work: Gallo, V., Chittka, L. 2018. "Cognitive aspects of comb-building in the honeybee?" *Frontiers in Psychology* 9: 900. DOI: 10.3389/fpsyg.2018.00900.

75: Comb construction influenced by the structure in which bees were raised: von Oelsen, G., Rademacher, E. 1979. "Untersuchungen zum Bauverhalten der Honigbiene (*Apis mellifica*)." *Apidologie* 10 (2): 175–209. DOI: 10.1051/apido:19790208.

78: Spider web construction requiring more than instinct: this was first expressed (with experimental support) by the then 25-year-old Charles Turner in 1892, when he wrote: "We may safely conclude that an instinctive impulse prompts gallery spiders to weave gallery webs, but the details of the construction are the products of intelligent action." Turner, C. H. 1892. "Psychological notes upon the gallery spider: illustrations of intelligent variations in the construction of the web." *Journal of Comparative Neurology* 2: 95–110. This idea has only recently gained traction again; see, e.g., Eberhard, W. G. 2019. "Adaptive flexibility in cues guiding spider web construction and its possible implications for spider cognition." *Behavior* 156 (3–4): 331–62. DOI: 10.1163/1568539X-00003544; Hesselberg, T. 2015. "Exploration behaviour and behavioural flexibility in orb-web spiders: a review." *Current Zoology* 61 (2): 313–27. DOI: 10.1093/czoolo/61.2.313.

78: Bees in space: Vandenberg, J. D., Massie, D. R., Shimanuki, H., Peterson, J. R., Poskevich, D. M. 1985. "Survival, behavior and comb construction by honeybees, *Apis mellifera*, in zero gravity aboard NASA shuttle mission STS-13." *Apidologie* 16 (4): 369–83. DOI: 10.1051/apido:19850402.

79: Simple explanations for seemingly clever animal behavior: Döring, T. F., Chittka, L. 2011. "How human are insects, and does it matter?" *Formosan Entomologist* 31: 85–99; Shettleworth, S. J. 2010. "Clever animals and killjoy explanations in comparative psychology." *Trends in Cognitive Sciences* 14 (11): 477–81. DOI: 10.1016/j.tics.2010.07.002.

79: Albrecht Bethe's work on bees' homing abilities: Bethe, A. 1898, *Dürfen wir den Ameisen und Bienen psychische Qualitäten zuschreiben?* Bonn: Verlag von Emil Strauss. The then-unidentified "homing sense" of hymenopteran insects had also been discussed by Jean-Henri Fabre and a correspondence with Charles Darwin on this topic, in volumes 1–2 of the *Souvenirs Entomologiques*.

80: Buttel-Reepen's responses to Bethe: Buttel-Reepen, H. 1900. "Sind die Bienen Reflexmaschinen?" *Experimentelle Beiträge zur Biologie der Honigbiene* 20: 1–84.

81: The controversy over flower syndromes, and the question of their rigidity: Clare, E. L., Schiestl, F. P., Leitch, A. R. , Chittka, L. 2013. "The promise of genomics in the study of plant-pollinator interactions." *Genome Biology* 14: 207. DOI: 10.1186/gb-2013-14-6-207; Fenster, C. B., Armbruster, W. S., Wilson, P., Dudash, M. R., Thomson, J. D. 2004. "Pollination syndromes and floral specialization." *Annual Review of Ecology, Evolution, and Systematics* 35: 375–403. DOI: 10.1146/annurev.ecolsys.34.011802.132347; Waser, N. M., Chittka, L., Price, M. V., Williams, N., Ollerton, J. 1996. "Generalization in pollination systems, and why it matters." *Ecology* 77: 1043–60. DOI: 10.2307/2265575. The latter reference contains the work in Strausberg, as well as evidence for flower specialists switching to other species when needed (see subsequent paragraphs).

83: Learning and instinct evolve together: Robinson, G. E., Barron, A. B. 2017. "Epigenetics and the evolution of instincts." *Science* 356 (6333): 26–27. DOI: 10.1126/science.aam6142.

84: Specialist bumble bees having to learn how to handle flowers: Laverty, T. M., Plowright, R. C. 1988, "Flower handling by bumblebees: a comparison of specialists and generalists." *Animal Behaviour* 36: 733–40. DOI: 10.1016/S0003-3472(88)80156-8.

**Chapter 5**

86: Hermann Müller quote: Müller, H. 1876. "Die Bedeutung der Honigbiene für unsere Blumen (IX)." *Bienenzeitung* 32: 176–84. Translated by the author.

86: 3D orientation at the root of human intelligence: Lorenz, K. 1978. *Behind the Mirror: A Search for a Natural History of Human Knowledge*. New York: Harcourt Brace Jovanovich.

87: For an overview of the long evolutionary history of insects: Grimaldi, D., Engel, M. S. 2005. *Evolution of the Insects*. Cambridge, UK: Cambridge University Press.

89: Wasps prospecting for future opportunities: van Nouhuys, S., Kaartinen, R. 2008. "A parasitoid wasp uses landmarks while monitoring potential resources." *Proceedings of the Royal Society B-Biological Sciences* 275 (1633): 377–85. DOI: 10.1098/rspb.2007.1446.

89: Brain and mushroom body evolution in hymenopteran insects: Farris, S. M., Schulmeister, S. 2011. "Parasitoidism, not sociality, is associated with the evolution of elaborate mushroom bodies in the brains of hymenopteran insects." *Proceedings of the Royal Society B-Biological Sciences* 278 (1707): 940–51. DOI: 10.1098/rspb.2010.2161; Godfrey, R. K., Gronenberg, W. 2019. "Brain evolution in social insects: advocating for the comparative approach." *Journal of Comparative Physiology A-Neuroethology, Sensory, Neural, and Behavioral Physiology* 205 (1): 13–32. DOI: 10.1007/s00359-019-01315-7; Sayol, F., Collado, M. A., Garcia-Porta, J., Seid, M. A., Gibbs, J., Agorreta, A., San Mauro, D., Raemakers, I., Sol, D., Bartomeus, I. 2020. "Feeding specialization and longer generation time are associated with relatively larger brains in bees." *Proceedings of the Royal Society B-Biological Sciences* 287 (1935). DOI: 10/1098/rspb.2020.0762.

90: Fabre on digger wasps: Fabre, J.-H. 1879. *Souvenirs Entomologiques—Ire série.* Paris, Charles Delagrave; for a very detailed account of the biology of one digger wasp species, see Baerends, G. P. 1941. "Fortpflanzungsverhalten und Orientierung der Grabwespe *Ammophila campestris*." *Tijdschrift voor Entomologie* 84: 71–248.

90: How one branch of wasps became vegan: see Grimaldi and Engel, *Evolution of the Insects*; or, for a popular scientific treatment, see Michael Engel quoted on p. 21 in Hanson, T. 2018. *Buzz.* New York: Basic Books; and Preston, C. 2006. *Bee.* London: Reaktion Books.

92: On the dance language of honey bees: see von Frisch, K. 1967. *The Dance Language and Orientation of Bees.* Cambridge, MA: Harvard University Press.

95: Martin Lindauer's work on Indian honey bees: Lindauer, M. 1956. "Über die Verständigung bei indischen Bienen." *Zeitschrift für Vergleichende Physiologie* 38: 521–57. DOI: 10.1007/BF00341108.

95: Further reading on the communication of Asian honey bees: Dyer, F. C. 1985. "Mechanisms of dance orientation in the Asian honey bee *Apis florea*." *Journal of Comparative Physiology A* 157: 183–98. DOI: 10.1007/BF01350026; Dyer, F. C. 1985. "Nocturnal orientation by the Asian honey bee, *Apis dorsata*." *Animal Behavior* 33: 769–74. DOI: 10.1016/S0003-3472(85)80009-9; Dyer, F. C. 1991. "Comparative studies of dance communication: analysis of phylogeny and function." In *Diversity in the Genus* Apis, ed. D. R. Smith, 177–98. Boulder, CO: Westview. DOI: 10.1201/9780429045868-9; Oldroyd, B. P., Wongsiri, S. 2006. *Asian Honey Bees—Biology, Conservation, and Human Interactions.* Cambridge, MA: Harvard University Press.

95: On the evolution of the honey bee dance language: Dyer, F. C. 2002. "The biology of the dance language." *Annual Review of Entomology* 47: 917–49. DOI: 10.1146/annurev.ento.47.091201.145306; Barron, A. B., Plath, J. A. 2017. "The evolution of honey bee dance communication: a mechanistic perspective." *Journal of Experimental Biology* 220 (23): 4339–46. DOI: 10.1242/jeb.142778.

96: "Why would bees follow a successful forager . . . ?" Some clues are provided by the (non-dancing) bumble bees, where uninformed individuals follow the movements of knowledgeable demonstrators very closely if the former individuals have reason to believe that a reward is to be had. See

Alem, S., Perry. C. J., Zhu, X., Loukola, O. J., Ingraham, T., Søvik, E., Chittka, L. 2016. "Associative mechanisms allow for social learning and cultural transmission of string pulling in an insect." *PLOS Biology* 14 (10): e1002564. DOI: 10.1371/journal.pbio.1002564. Honey bee dancers provide these rewards during trophallaxis, in which they regurgitate food that is consumed by dance attendees. Thus, sugar rewards may have facilitated the following behavior that is crucial for dance communication to work.

97: Martin Lindauer's work with Warwick Kerr on stingless bees: Lindauer, M., Kerr, W. 1958. "Die gegenseitige Verständigung bei den stachellosen Bienen." *Zeitschrift für Vergleichende Physiologie* 41: 405–34. DOI: 10.1007/BF00344263; for a more recent overview of the diverse biology of stingless bees and their communication systems, see: Grüter, C. 2020. *Stingless Bees*. Berlin: Springer Verlag,

97: Vibrational pulses correlated in length to the distance of a food source: see Grüter, *Stingless Bees*, and Nieh, J. C. 2004. "Recruitment communication in stingless bees (Hymenoptera, Apidae, Meliponini)." *Apidologie* 35 (2): 159–82. DOI: 10.1051/apido:2004007.

97: Intention movements of honey bees: Dyer, "The biology of the dance language."

98: Stingless bees are thought to be the sister group of bumble bees: Romiguier, J., Cameron, S. A., Woodard, S. H., Fischman, B. J., Keller, L., Praz, C. J. 2016. "Phylogenomics controlling for base compositional bias reveals a single origin of eusociality in corbiculate bees." *Molecular Biology and Evolution* 33 (3): 670–78. DOI: 10.1093/molbev/msv258.

98: Anna Dornhaus's work on the communication of bumble bees: Dornhaus, A., Chittka, L. 1999. "Evolutionary origins of bee dances." *Nature* 401: 38–38. DOI: 10.1038/43372; Dornhaus, A., Chittka, L. 2001. "Food alert in bumblebees: possible mechanisms and evolutionary implications." *Behavioral Ecology and Sociobiology* 50: 570–76. DOI: 10/1007/s002650100395; Dornhaus, A., Brockmann, A., Chittka, L. 2003. "Bumble bees alert to food with pheromone from tergal gland." *Journal of Comparative Physiology A* 189: 47–51. DOI: 10.1007/s00359-002-0374-y.

99: Anna Dornhaus's work on the adaptive significance of the honey bee dance language: Dornhaus, A., Chittka, L. 2004. "Why do honeybees dance?" *Behavioural Ecology and Sociobiology* 55: 395–401. DOI: 10.1007/s00265-003-0726-9. We originally submitted this work to the journal *Nature* in 2001, but it was rejected by the editor without review. Curiously, the same journal later published a study using the same methods and drawing similar conclusions, but by different authors (Sherman, G., Visscher, P. K. 2002. "Honeybee colonies achieve fitness through dancing." *Nature* 419: 920–22. DOI: 10.1038/nature01127). The world of scientific publication is curious and haphazard sometimes.

**Chapter 6**

102: Dzierzon quote: Dzierzon, J. Letter to Hugo von Buttel-Reepen. Quoted in Buttel-Reepen, H. 1900. "Sind die Bienen Reflexmaschinen?" *Experimen-*

*telle Beiträge zur Biologie der Honigbiene* 20: 1–84. Translated by the author.

102: Fabre's experiments on homing solitary bees and wasps, and correspondence with Darwin on the topic, are in volumes 1–2 of the *Souvenirs Entomologiques*.

103: Citations for the biographical details of Charles Turner are in the references for chapter 1.

104: Charles Turner's Coca Cola cap experiment is described in: Turner, C. H. 1908. "The homing of the burrowing-bees (Anthrophodidae)." *Biological Bulletin* 15: 247–58; similar experiments were performed much later by Niko Tinbergen on beewolf digger wasps. The Nobel laureate (awarded for studies of individual and social behavior in animals) first marked a beewolf's nest entrance with pine cones, then moved them to demonstrate that the insect was guided by the memory of the landmarks. Tinbergen, N. 1932. "Über die Orientierung des Bienenwolfes." *Zeitschrift für Vergleichende Physiologie* 16: 305–34. Unfortunately the world now largely credits Tinbergen, not Turner, with this discovery. I encourage readers to explore Turner's original writings—they are a gold mine of scientific ideas, and also very engagingly written and at times poetic. A taster: "The time was the month of August; the place, an abandoned garden in Augusta, Georgia. In one end of that garden, where a poor stand of grass had supplanted the beans which once flourished there, a host of burrowing-bees, bearing the generic name *Melissodes*, had excavated their burrows. Some of these burrows were exposed in barren spots, while others were hidden, more or less, in clumps of grass. At any hour of the day, these industrious burrowing-bees could be seen storing their nests with pollen; while the cuckoo-bees . . . loitered on neighboring grass-blades or hovered about the burrows, waiting for an opportunity to lay their eggs upon the stored-up food of the industrious *Melissodes*. Crossing the path of the breeze and the rays of the sun at every possible angle, active alike in sunshine and shadow, yet disturbed by even slight changes in the topographical environment of the burrow, the female *Melissodes* continues her work from early morn until set of sun; proclaiming, in deeds more eloquent than words: 'My behavior is much more than a complex of anemotropisms and phototropisms, for my homing is controlled by memory pictures of the environment of my nest.'" Turner, C. H. 1908. "The sun-dance of Melissodes." *Psyche* 15: 122–24. DOI: 10.1155/1908/632919. Further writings by Turner on hymenopteran orientation: 1912. "Sphex overcoming obstacles." *Psyche* 19: 100–101. DOI: 10.1155/1912/95842; 1923. "The homing of the Hymenoptera." *Transactions of the Academy of Science of St. Louis* 24: 27–45.

105: Context learning in bees: Collett, T. S., Kelber, A. 1988. "The retrieval of visuo-spatial memories by honeybees." *Journal of Comparative Physiology A* 163: 145–50. DOI: 10.1007/BF00612004.

105: Further studies on context learning in bumble bees and honey bees: Collett, T. S., Fauria, K., Dale, K., Baron, J. 1997. "Places and patterns—a study of context learning in honeybees." *Journal of Comparative Physiology A* 181: 343–53. DOI: 10.1007/s003590050120; Fauria, K., Dale, K., Colborn, M., Collett, T. S. 2002. "Learning speed and contextual isolation in bumblebees." *Journal of Experimental Biology* 205 (7): 1009–18. DOI: 10.1242/jeb.205.7.1009.

105: Illumination as a contextual cue: Lotto, R. B., Chittka, L. 2005. "Seeing the light: Illumination as a contextual cue to color choice behavior in bumblebees." *Proceedings of the National Academy of Sciences of the USA* 102: 3852–56. DOI: 10.1073/pnas.0500681102.

106: Original study on cognitive maps in bees: Gould, J. L. 1986. "The locale map of honey bees: Do insects have cognitive maps?" *Science* 232: 861–63. DOI: 10.1126/science.232.4752.861; a good overview on critical tests of cognitive maps is Bennett, A.T.D. 1996. "Do animals have cognitive maps?" *The Journal of Experimental Biology* 199: 219–24. DOI: 10.1242/jeb.199.1.219.

107: Martin Lindauer's work on nocturnal dances: Lindauer, M. 1954. "Dauertänze im Bienenstock und ihre Beziehung zur Sonnenbahn." *Naturwissenschaften* 41: 506–7. DOI: 10.1007/BF00631843; and further details in von Frisch, K. *The Dance Language and Orientation of Bees*. 1967. Cambridge, MA: Harvard University Press; for a cognitive interpretation, see Menzel, R., Eckoldt, M. 2016. *Die Intelligenz der Bienen*. München: Knaus.

107: Recruits rejecting dances indicating a location in a lake: Gould, J. L., Gould, C. G. 1982. "The insect mind—physics or metaphysics?" In *Animal Mind—Human Mind*, ed. D. R. Griffin, 269–98. Berlin: Springer Verlag; but see the refutation: Wray, M. K., Klein, B. A., Mattila, H. R., Seeley, T. D. 2008. "Honeybees do not reject dances for 'implausible' locations: reconsidering the evidence for cognitive maps in insects." *Animal Behaviour* 76: 261–69. DOI: 10.1016/j.anbehav.2008.04.005.

109: Early experimental evidence against cognitive map use in bees: Menzel, R., Chittka, L., Eichmüller, S., Geiger, K., Peitsch, D., Knoll, P. 1990. "Dominance of celestial cues over landmarks disproves map-like orientation in honey bees." *Zeitschrift für Naturforschung C* 45 (6): 723–26. DOI: 10.1515/znc-1990-0625; Wehner, R., Bleuler, S., Nievergelt, C., Shah, D. 1990. "Bees navigate by using vectors and routes rather than maps." *Naturwissenschaften* 77 (10): 479–82. DOI: 10.1007/bf01135926; Dyer, F. C. 1991. "Bees acquire route-based memories but not cognitive maps in a familiar landscape." *Animal Behaviour* 41: 239–46. DOI: 10.1016/S0003-3472(05)80475-0.

111: Sun compass versus local landmarks in direction finding in a featureless landscape: Chittka, L., Geiger, K. 1995. "Honeybee long-distance orientation in a controlled environment." *Ethology* 99: 117–26. DOI: 10.1111/j.1439-0310.1995.tb01093.x.

113: Study on landmark counting in honey bees: Chittka, L., Geiger, K. 1995. "Can honeybees count landmarks?" *Animal Behaviour* 49: 159–64. DOI: 10.1016/0003-3472(95)80163-4.

113: Further verifications of numerical abilities in various species of bees:

Dacke, M., Srinivasan, M. V. 2008. "Evidence for counting in insects." *Animal Cognition* 11: 683–89. DOI: 10.1007/s10071-008-0159-y; Gross, H. J., Pahl, M., Si, A., Zhu, H., Tautz, J., Zhang, S. 2009. "Number-based visual generalisation in the honeybee. *PLOS One* 4: e4263. DOI: 10.1371/journal.pone.0004263; Pahl, M., Si, A., Zhang, S. 2013. "Numerical cognition in bees and other insects." *Frontiers in Psychology* 4: 162. DOI: 10.3389/fpsyg.2013.00162; Bar-Shai, N., Keasar, T., Shmida, A. 2011. "The use of numerical information by bees in foraging tasks." *Behavioral Ecology* 22: 317–25. DOI: 10.1093/beheco/arq206; Bar-Shai, N., Keasar, T., Shmida, A. 2011. "How do solitary bees forage in patches with a fixed number of food items?" *Animal Behaviour* 82: 1367–72. DOI: 10.1016/j.anbehav.2011.09.020. There was also a recent flurry of studies on more-advanced numerical abilities of bees, purporting to show that bees can add and subtract, and even understand the concept of zero; see Howard, S. R., Avargues-Weber, A., Garcia, J. E., Greentree, A. D., Dyer, A. G. 2018. "Numerical ordering of zero in honey bees." *Science* 360: 1124–26. DOI: 10.1126/science.aar4975; Howard, S. R., Avargues-Weber, A., Garcia, J. E., Greentree, A. D., Dyer, A. G. 2019. "Numerical cognition in honeybees enables addition and subtraction." *Science Advances* 5 (2). DOI: 10.1126/sciadv.aav0961. However, it is presently not fully clear whether bees used number or some alternative cue to solve these tasks: MaBouDi, H., Barron, A. B., Li, S., Honkanen, M., Loukola, O. J., Peng, F., Li, W., Marshall, J.A.R., Cope, A., Vasilaki, E., Solvi, C. 2021. "Non-numerical strategies used by bees to solve numerical cognition tasks." *Proceedings of the Royal Society B—Biological Sciences* 288: 20202711. DOI: 10.1098/rspb.2020.2711.

114: Sequential nature of counting in bees: Skorupski, P., MaBouDi, H., Galpayage, Dona H. S., Chittka, L. 2018. "Counting insects." *Philosophical Transactions of the Royal Society B—Biological Sciences* 373: 20160513. DOI: 10.1098/rstb.2016.0513; MaBouDi, H., Galpayage, Dona H. S., Gatto, E., Loukola, O. J., Buckley, E., Onoufriou, P. D., Skorupski, P., Chittka, L. 2020. "Bumblebees use sequential scanning of countable items in visual patterns to solve numerosity tasks." *Integrative and Comparative Biology* 60: 929–42. DOI: 10.1093/icb/icaa025.

115: Path integration in desert ants: Müller, M., Wehner, R. 1988. "Path integration in desert ants, *Cataglyphis fortis*." *Proceedings of the National Academy of Sciences of the USA* 85: 5287–90. DOI: 10.1073/pnas.85.14.5287; Collett, T. S., Collett, M. 2000. "Path integration in insects." *Current Opinion in Neurobiology* 10: 757–62. DOI: 10.1016/s0959-4388(00)00150-1; Collett, M., Collett, T. S. 2017. "Path integration: combining optic flow with compass orientation." *Current Biology* 27 (20): R1113–16. DOI: 10.1016/j.cub.2017.09.004.

115: Von Frisch's evidence for path integration in dancing bees: von Frisch, *The Dance Language*; further explained in: Collett, M., Collett, T. S. 2000. "How do insects use path integration for their navigation?" *Biological Cybernetics* 83: 245–59. DOI: 10.1007/s004220000168.

117: Path integration in honey bees in an Arizona desert: Chittka, L., Kunze, J., Shipman, C., Buchmann, S. L. 1995. "The significance of landmarks for path integration of homing honey bee foragers." *Naturwissenschaften* 82: 341–43. DOI: 10.1007/BF01131533.

118: Use of optic flow in measuring flight distance: Srinivasan, M. V., Zhang, S., Altwein, M., Tautz, J. 2000. "Honey-bee navigation: nature and calibration of the 'odometer.'" *Science* 287: 851–53. DOI: 10.1126/science.287.5454.851; Esch, H. E., Zhang, S., Srinivasan, M. V., Tautz, J. 2001. "Honeybee dances communicate distances measured by optic flow." *Nature* 411: 581–83. DOI: 10.1038/35079072; Tautz, J., Zhang, S., Spaethe, J., Brockmann, A., Si, A., Srinivasan, M. V. 2004. "Honeybee odometry: performance in varying natural terrain." *PLOS Biology* 2: e211. DOI: 10.1371/journal.pbio.0020211; Chittka, L. 2004. "Dances as windows into insect perception." *PLOS Biology* 2: 898–900. DOI: 10.1371/journal.pbio.0020216.

118: Path integration fails in the absence of visual cues in bumble bees: Chittka, L., Williams, N., Rasmussen, H., Thomson, J. D. 1999. "Navigation without vision—bumble bee orientation in complete darkness." *Proceedings of the Royal Society B—Biological Sciences* 266: 45–50. DOI: 10.1098/rspb.1999.0602.

119: Neural models of path integration: Stone, T., Webb, B., Adden, A., Ben Weddig, N., Honkanen, A., Templin, R., Wcislo, W., Scimeca, L., Warrant, E., Heinze, S. 2017. "An anatomically constrained model for path integration in the bee brain." *Current Biology* 27 (20): 3069–85. DOI: 10.1016/j.cub.2017.08.052.

120: First study on radar tracking of bees: Riley, J. R., Smith, A. D., Reynolds, D. R., Edwards, A. S., Osborne, J. L., Williams, I. H., Carreck, N. L., Poppy, G. M. 1996. "Tracking bees with harmonic radar." *Nature* 379: 29–30. DOI: 10.1038/379029b0.

120: Lifetime radar tracking of bees: Woodgate, J. L., Makinson, J. C., Lim, K. S., Reynolds, A. M., Chittka, L. 2016. "Life-long radar tracking of bumblebees." *PLOS One* 11 (8): 22. DOI: 10.1371/journal.pone.0160333. Obtaining such histories is obviously easier for walking than for flying animals, and so such lifetime tracking was performed much earlier in ants: Wehner, R., Harkness, R. D., Schmid-Hempel, P. 1983. "Foraging strategies in individually searching ants, *Cataglyphis bicolor* (Hymenoptera: Formicidae)." In *Information Processing in Animals*, ed. M. Lindauer, 1–79. Stuttgart: Gustav Fischer Verlag.

124: Randolf Menzel supporting the notion of cognitive maps in bees: Menzel, R., Greggers, U., Smith, A., Berger, S., Brandt, R., Brunke, S., Bundrock, G., Hulse, S., Plumpe, T., Schaupp, F., et al. 2005. "Honey bees navigate according to a map-like spatial memory." *Proceedings of the National Academy of Sciences of the USA* 102 (8): 3040–45. DOI: 10.1073/pnas.0408550102.

124: Experiments with jet-lagged bees: Cheeseman, J. F., Millar, C. D., Greggers, U., Lehmann, K., Pawley, M.D.M., Gallistel, C. R., Warman,

G. R., Menzel, R. 2014. "Way-finding in displaced clock-shifted bees proves bees use a cognitive map." *Proceedings of the National Academy of Sciences of the USA* 111 (24): 8949–54. DOI: 10.1073/pnas.1408039111.

126: Criticism of the study of jet-lagged bees: Cheung, A., Collett, M., Collett, T. S., Dewar, A., Dyer, F., Graham, P., Mangan, M., Narendra, A., Philippides, A., Sturzl, W., et al. 2014. "Still no convincing evidence for cognitive map use by honeybees." *Proceedings of the National Academy of Sciences of the USA* 111 (42): E4396–97. DOI: 10.1073/pnas.1413581111.

126: Smells reactivating spatial memories in bees: Reinhard, J., Srinivasan, M. V., Guez, D., Zhang, S. W. 2004. "Floral scents induce recall of navigational and visual memories in honeybees." *Journal of Experimental Biology* 207 (25): 4371–81. DOI: 10.1242/jeb.01306.

127: James Thomson's work on traplining bees: Thomson, J. D., Peterson, S. C., Harder, L. D. 1987. "Response of traplining bumble bees to competition experiments: shifts in feeding location and efficiency." *Oecologia* 71: 295–300. DOI: 10.1007/BF00377298; Thomson, J. D. 1996. "Trapline foraging by bumblebees: I. Persistence of flight-path geometry." *Behavioral Ecology* 7 (2): 158–64. DOI: 10.1093/beheco/7.2.158; Thomson, J. D., Slatkin, M., Thomson, B. A. 1997. "Trapline foraging by bumble bees: II. Definition and detection from sequence data." *Behavioral Ecology* 8 (2): 199–210. DOI: 10.1093/beheco/8.2.199; Williams, N. M., Thomson, J. D. 1998. "Trapline foraging by bumble bees: III. Temporal patterns of visitation and foraging success at single plants." *Behavioral Ecology* 9 (6): 612–21. DOI: 10.1093/beheco/9.6.612.

127: Author's interest in sequence learning in bees: Chittka, L., Kunze, J., Geiger, K. 1995. "The influences of landmarks on distance estimation of honeybees." *Animal Behaviour* 50: 23–31. DOI: 10.1006/anbe.1995.0217.

128: Radar tracking of bees in a traplining study: Lihoreau, M., Raine, N. E., Reynolds, A. M., Stelzer, R. J., Lim, K. S., Smith, A.D., Osborne J. L., Chittka L. 2012. "Radar tracking and motion-sensitive cameras on flowers reveal the development of pollinator multi-destination routes over large spatial scales." *PLOS Biology* 10 (9). DOI: 10.1371/journal.pbio.1001392.

129: Further overviews of insect spatial learning: Collett, M., Chittka, L., Collett, T. S. 2013. "Spatial memory in insect navigation." *Current Biology* 23 (17): R789–800. DOI: 10.1016/j.cub.2013.07.020; Srinivasan, M. V. 2011. "Honeybees as a model for the study of visually guided flight, navigation, and biologically inspired robotics." *Physiological Reviews* 91 (2): 413–60. DOI: 10.1152/physrev.00005.2010; Cruse, H., Wehner, R. 2011. "No need for a cognitive map: decentralized memory for insect navigation." *PLOS Computational Biology* 7 (3). DOI: 10.1371/journal.pcbi.1002009.

**Chapter 7**

130: Darwin quote: Darwin, C. 1876. *The Effects of Cross and Self Fertilization in the Vegetable Kingdom*. London: John Murray.

131: Learning to handle electronic flowers: Chittka, L., Thomson, J. D. 1997.

"Sensori-motor learning and its relevance for task specialization in bumble bees." *Behavioral Ecology and Sociobiology* 41: 385–98. DOI: 10.1007/s002650050400; Chittka, L. 1998. "Sensorimotor learning in bumblebees: long term retention and reversal training." *Journal of Experimental Biology* 201: 515–24. DOI: 10.1242/jeb.201.4.515; Chittka, L. 2002. "The influence of intermittent rewards on learning to handle flowers in bumblebees." *Entomologia Generalis* 26: 85–91.

135: Overviews of the biology of attention in animals: Dukas, R. 2004. "Causes and consequences of limited attention." *Brain Behavior and Evolution* 63 (4): 197–210. DOI: 10.1159/000076781; Nityananda, V. 2016. "Attention-like processes in insects." *Proceedings of the Royal Society B—Biological Sciences* 283 (1842). DOI: 10.1098/rspb.2016.1986.

137: Behavioral limits of flower detection in honey bees: Lehrer, M., Bischof, S. 1995. "Detection of model flowers by honeybees: the role of chromatic and achromatic contrast." *Naturwissenschaften* 82: 145–47. DOI: 10.1007/BF01177278; Giurfa, M., Vorobyev, M., Kevan, P., Menzel, R. 1996. "Detection of coloured stimuli by honeybees: minimum visual angles and receptor specific contrasts." *Journal of Comparative Physiology A* 178: 699–709. DOI: 10.1007/BF00227381.

137: On the limits of spatial resolution in insect eyes, more-recent evidence indicates that resolution might be higher than the simple optics of the ommatidial array suggests. Through photoreceptor contraction and small-scale eye movements (saccades), insects might achieve far better resolution than previously thought: Juusola, M., Dau, A., Song, Z. Y., Solanki, N., Rien, D., Jaciuch, D., Dongre, S., Blanchard, F., de Polavieja, G. G., Hardie, R. C., Jouni, T. 2017. "Microsaccadic sampling of moving image information provides *Drosophila* hyperacute vision." *eLife* 6. DOI: 10.7554/eLife.26117.

137: Two processing channels for visual information (with different speeds) in the bumble bee: Spaethe, J., Tautz, J., Chittka, L. 2001. "Visual constraints in foraging bumblebees: flower size and color affect search time and flight behavior." *Proceedings of the National Academy of Sciences of the USA* 98 (7): 3898–903. DOI: 10.1073/pnas.071053098.

139: Serial search in honey bees: Spaethe, J., Tautz, J., Chittka, L. 2006. "Do honeybees detect colour targets using serial or parallel visual search?" *Journal of Experimental Biology* 209: 987–93. DOI: 10.1242/jeb.02124; comparison with bumblebees: Morawetz, L., Spaethe, J. 2012. "Visual attention in a complex search task differs between honeybees and bumblebees." *Journal of Experimental Biology* 215 (14): 2515–23. DOI: 10.1242/jeb.066399.

139: Vivek Nityananda's work on what bumble bees can and cannot see in brief stimulus presentations: Nityananda, V., Skorupski, P., Chittka, L. 2014. "Can bees see at a glance?" *Journal of Experimental Biology* 217 (11): 1933–39. DOI: 10.1242/jeb.101394.

141: Speed-accuracy tradeoffs in insect decision making were first hinted at in a study by Charles Turner: Turner,

C. H. 1913. "Behavior of the common roach (*Periplaneta orientalis* L.) on an open maze." *Biological Bulletin* 25: 348–65. Our work on bees is in: Chittka, L., Dyer, A. G., Bock, F., Dornhaus, A. 2003. "Bees trade off foraging speed for accuracy." *Nature* 424: 388. DOI: 10.1038/424388a; a broader overview about such tradeoffs, and their reasons and implications, is: Chittka, L., Skorupski, P., Raine, N. E. 2009. "Speed-accuracy tradeoffs in animal decision making." *Trends in Ecology & Evolution* 24: 400–407. DOI: 10.1016/j.tree.2009.02.010.

143: Honey bees recognizing images of human faces: Dyer, A. G., Neumeyer, C., Chittka, L. 2005. "Honeybee (*Apis mellifera*) vision can discriminate between and recognise images of human faces." *Journal of Experimental Biology* 208 (24): 4709–14. DOI: 10.1242/jeb.01929; for further information on the psychological mechanisms involved in processing such stimuli in bees: Avargues-Weber, A., Portelli, G., Benard, J., Dyer, A. G., Giurfa, M. 2010. "Configural processing enables discrimination and categorization of face-like stimuli in honeybees." *Journal of Experimental Biology* 213 (4): 593–601. DOI: 10.1242/jeb.039263.

143: On face recognition in humans: Kanwisher, N. 2000. "Domain-specificity in face perception." *Nature Neuroscience* 3: 759–63. DOI: 10.1038/77664; Tsao, D. Y., Freiwald, W. A., Tootell, R.B.H., Livingstone, M. S. 2006. "A cortical region consisting entirely of face-selective cells." *Science* 311: 670–74. DOI: 10.1126/science.1119983.

143: Face recognition in wasps: Sheehan, M. J., Tibbetts, E. A. 2011. "Specialized face learning is associated with individual recognition in paper wasps." *Science* 334 (6060): 1272–75. DOI: 10.1126/science.1211334—and a popular scientific account of this study: Chittka, L., Dyer, A. 2012. "Cognition: your face looks familiar." *Nature* 481 (7380): 154–55. DOI: 10.1038/481154a.

144: Bee interactions with flower texture: Whitney, H. M., Chittka, L., Bruce, T.J.A., Glover, B. J. 2009. "Conical epidermal cells allow bees to grip flowers and increase foraging efficiency." *Current Biology* 19: 948–53. DOI: 10.1016/j.cub.2009.04.051; Whitney, H. M., Bennet, K.M.V., Dorling, M., Sandbach, L., Prince, D., Chittka, L., Glover, B. J. 2011. "Why do so many petals have conical epidermal cells?" *Annals of Botany* 108 (4): 609–16. DOI: 10.1093/aob/mcr065. Already in the 1980s, Canadian scientists Peter Kevan and Meredith Lane experimented with gold-coated flower petals and discovered that bees could use their antennae to discriminate microstructures of petals.

145: On thermoregulation in insects, see writings by Bernd Heinrich: Heinrich, B. 1993. *The Hot-Blooded Insects: Strategies and Mechanisms of Thermoregulation*. Berlin: Springer Verlag; Heinrich, B., Esch, H. 1994. "Thermoregulation in bees." *American Scientist* 82 (2): 164–70; Heinrich, B. 1996. *The Thermal Warriors—Strategies of Insect Survival*. Cambridge, MA: Harvard University Press.

146: Bees learning to visit warmer flowers: Dyer, A. G., Whitney, H. M., Arnold, S.E.J., Glover, B. J., Chittka, L. 2006.

"Bees associate warmth with floral colour." *Nature* 442 (7102): 525. DOI: 10.1038/442525a; Whitney, H. M., Dyer, A., Chittka, L., Rands, S. A., Glover, B. J. 2008. "The interaction of temperature and sucrose concentration on foraging preferences in bumblebees." *Naturwissenschaften* 95: 845–50. DOI: 10.1007/s00114-008-0393-9; a popular scientific account is in: Whitney, H., Chittka, L. 2007. "Warm flowers, happy pollinators." *Biologist* 54: 154–59.

147: Bees learning iridescence as a floral cue: Whitney, H. M., Kolle, M., Andrew, P., Chittka, L., Steiner, U., Glover, B. J. 2009. "Floral iridescence, produced by diffractive optics, acts as a cue for animal pollinators." *Science* 323 (5910): 130–33. DOI: 10.1126/science.1166256.

148: Iridescence increasing flower detectability: Whitney, H. M., Reed, A., Rands, S. A., Chittka, L., Glover, B. J. 2016. "Flower iridescence increases object detection in the insect visual system without compromising object identity." *Current Biology* 26 (6): 802–8. DOI: 10.1016/j.cub.2016.01.026.

148: Bees learning polarization patterns on floral targets: Foster, J. J., Sharkey, C. R., Gaworska, A.V.A., Roberts, N. W., Whitney, H. M., Partridge, J. C. 2014. "Bumblebees learn polarization patterns." *Current Biology* 24 (12): 1415–20. DOI: 10.1016/j.cub.2014.05.007.

148: Honey bees' preference for flower colors in the violet-blue range: Giurfa, M., Nunez, J., Chittka, L., Menzel, R. 1995. "Colour preferences of flower-naive honeybees." *Journal of Comparative Physiology A* 177: 247–59. DOI: 10.1007/BF00192415; for a variety of bumble bee species and populations: Raine, N. E., Ings, T. C., Dornhaus, A., Saleh, N., Chittka, L. 2006. "Adaptation, genetic drift, pleiotropy, and history in the evolution of bee foraging behavior." *Advances in the Study of Behavior* 36: 305–54. DOI: 10.1016/S0065-3454(06)36007-X; Raine, N. E., Chittka, L. 2007. "The adaptive significance of sensory bias in a foraging context: floral colour preferences in the bumblebee *Bombus terrestris*." *PLOS One* 2: e556. DOI: 10.1371/journal.pone.0000556.

149: Bees learning to classify flowers based on symmetry: Giurfa, M., Eichmann, B., Menzel, R. 1996. "Symmetry perception in an insect." *Nature* 382: 458–61. DOI: 10.1038/382458a0.

150: Bees learning patterns of sameness and difference: Giurfa, M., Zhang, S., Jenett, A., Menzel, R., Srinivasan, M. V. 2001. "The concepts of 'sameness' and 'difference' in an insect." *Nature* 410: 930–33. DOI: 10.1038/35073582.

152: Christof Koch on bee consciousness: Koch, C. 2008. "Exploring consciousness through the study of bees." *Scientific American*, December 1, 2008. http://www.scientificamerican.com/article/exploring-consciousness/.

152: Bees learning about intervals between rewards: Boisvert, M. J., Sherry, D. F. 2006. "Interval timing by an invertebrate, the bumble bee *Bombus impatiens*." *Current Biology* 16 (16): 1636–40. DOI: 10.1016/j.cub.2006.06.064; for a popular scientific account, see Skorupski, P., Chittka, L. 2006 "Animal cognition: an insect's sense of time?" *Current Biology* 16 (19): R851–53. DOI: 10.1016/j.cub.2006.08.069.

152: Learning to withhold responses for a given time: Shamosh, N. A., DeYoung, C. G., Green, A. E., Reis, D. L., Johnson, M. R., Conway, A.R.A., Engle, R. W., Braver, T. S., Gray, J. R. 2008. "Individual differences in delay discounting relation to intelligence, working memory, and anterior prefrontal cortex." *Psychological Science* 19 (9): 904–11. DOI: 10.1111/j.1467-9280.2008.02175.x; and also MacLean, E. L., Hare, B., Nunn, C. L., Addessi, E., Amici, F., Anderson, R. C., Aureli, F., Baker, J. M., Bania, A. E., Barnard, A. M., et al. 2014. "The evolution of self-control." *Proceedings of the National Academy of Sciences of the USA* 111 (20): E2140–48. DOI: 10.1073/pnas.1323533111.

153: On the usefulness of interval timing in flower visitors: Williams, N. M., Thomson, J. D. 1998. "Trapline foraging by bumble bees: III. Temporal patterns of visitation and foraging success at single plants." *Behavioral Ecology* 9 (6): 612–21. DOI: 10.1093/beheco/9.6.612.

153: On the suggestion that bees can form spatial concepts: Avargues-Weber, A., Dyer, A. G., Giurfa, M. 2011. "Conceptualization of above and below relationships by an insect." *Proceedings of the Royal Society B—Biological Sciences* 278 (1707): 898–905. DOI: 10.1098/rspb.2010.1891; for a popular scientific account, see: Chittka, L., Jensen, K. 2011. "Animal cognition: concepts from apes to bees." *Current Biology* 21 (3): R116–19. DOI: 10.1016/j.cub.2010.12.045.

155: An alternative explanation for how bees might solve a spatial concept learning task: Guiraud, M., Roper, M., Chittka, L. 2018. "High-speed videography reveals how honeybees can turn a spatial concept learning task into a simple discrimination task by stereotyped flight movements and sequential inspection of pattern elements." *Frontiers in Psychology* 9: 1347. DOI: 10.3389/fpsyg.2018.01347.

**Chapter 8**

158: Darwin quotes from: Romanes, G. J. 1884. *Mental Evolution in Animals.* New York: AMS Press; Darwin, C. R. 1841. Letter no. 607, from Charles Darwin to *The Gardeners' Chronicle,* August 21, 1841. In *The Correspondence of Charles Darwin,* vol. 2, 1837–43. Cambridge: Cambridge University Press, 1986.

160: Bees learning by observation: Leadbeater, E., Chittka, L. 2005. "A new mode of information transfer in foraging bumblebees?" *Current Biology* 15 (12): R447–48. DOI: 10.1016/j.cub.2005.06.011.

160: University of Arizona study on bees learning by observing other bees through a screen: Worden, B. D., Papaj, D. R. 2005. "Flower choice copying in bumblebees." *Biology Letters* 1: 504–7. DOI: 10.1098/rsbl.2005.0368.

161: Second-order conditioning in bumble bee social learning: Dawson, E. H., Avargues-Weber, A., Chittka, L., Leadbeater, E. 2013. "Learning by observation emerges from simple associations in an insect model." *Current Biology* 23 (8): 727–30. DOI: 10.1016/j.cub.2013.03.035.

163: Bees learning about "warning lights": Dawson, E. H., Chittka, L., Leadbeater, E. 2016. "Alarm substances induce associative social learning in honeybees, *Apis mellifera.*"

*Animal Behaviour* 122: 17–22. DOI: 10.1016/j.anbehav.2016.08.006.

163: Bees learning from other species: Dawson, E. H., Chittka, L. 2012. "Conspecific and heterospecific information use in bumble bees." *PLOS One* 7 (2): e31444. DOI: 10.1371/journal.pone.0031444; Romero Gonzalez, E. R., Solvi, C., Chittka, L. 2020. "Honeybees adjust colour preferences in response to concurrent social information from conspecifics and heterospecifics." *Animal Behaviour* 170: 219–28. DOI: 10.1016/j.anbehav.2020.10.008.

163: Honey bees responding accurately to another species' dance dialect: Su, S., Cai, F., Si, A., Zhang, S., Tautz, J., Chen, S. 2008. "East learns from west: Asiatic honey bees can understand dance language of european honey bees." *PLOS One* 3 (6): 1–9. DOI: 10.1371/journal.pone.0002365.

164: Heterospecific social learning in stingless bees: Nieh, J. C., Barreto, L. S., Contrera, F.A.L., Imperatriz-Fonseca, V. L. 2004. "Olfactory eavesdropping by a competitively foraging stingless bee, *Trigona spinipes*." *Proceedings of the Royal Society B—Biological Sciences* 271 (1548): 1633–40. DOI: 10.1098/rspb.2004.2717.

164: Learning tricks of thievery: Leadbeater, E., Chittka, L. 2008. "Social transmission of nectar-robbing behaviour in bumble-bees." *Proceedings of the Royal Society B—Biological Sciences* 275 (1643): 1669–74. DOI: 10.1098/rspb.2008.0270.

166: Nectar robbing in Alpine bumble bees: Goulson, D., Park, K. J., Tinsley, M. C., Bussière, L. F., Vallejo-Marin, M. 2013. "Social learning drives handedness in nectar-robbing bumblebees." *Behavioral Ecology and Sociobiology* 67 (7): 1141–50. DOI: 10.1007/s00265-013-1539-0.

166: Traditions in honey bees: Lindauer, M. 1985. "The dance language of honeybees: the history of a discovery." In *Experimental Behavioral Ecology*, eds. B. Hölldobler, M. Lindauer, 129–40. Stuttgart: G. Fischer Verlag; see also: Kirchner, W. H. 1987. "Tradition im Bienenstaat. Kommunikation zwischen Imagines und der Brut der Honigbiene durch Vibrationssignale." PhD Thesis, University of Würzburg.

169: String pulling in bumble bees: Alem, S., Perry, C. J., Zhu, X. F., Loukola, O. J., Ingraham, T., Sovik, E., Chittka, L. 2016. "Associative mechanisms allow for social learning and cultural transmission of string pulling in an insect." *PLOS Biology* 14 (10): e1002564. DOI: 10.1371/journal.pbio.1002564. (Note: The second author of the study is now named Cwyn Solvi.)

170: Bees learning better from moving than stationary individuals: Avargues-Weber, A., Chittka, L. 2014. "Observational conditioning in flower choice copying by bumblebees (*Bombus terrestris*): influence of observer distance and demonstrator movement." *PLOS One* 9 (2): e88415. DOI: 10.1371/journal.pone.0088415.

171: Bees learning from each other to roll balls: Loukola, O. J., Solvi, C., Coscos, L., Chittka, L. 2017. "Bumblebees show cognitive flexibility by improving on an observed complex behavior." *Science* 355 (6327): 833–36. DOI: 10.1126/science.aag2360.

173: Charles Turner's work suggesting outcome awareness in insects and other animals: Turner, C. H. 1907. "Do ants form practical judgments?" *Biological Bulletin* 13: 333–43. DOI: 10.2307/1535609; Turner, C. H. 1909. "Behavior of a snake." *Science* 30: 563–64. DOI: 10.1126/science.30.773.563.

174: For a comprehensive overview of the bee swarming process, see: Seeley, T. D. 2010. *Honeybee Democracy*. Princeton: Princeton University Press, and references therein.

175: Maeterlinck quote: Maeterlinck, M. 1901. *La vie des abeilles*. Paris, Editions Frasquelle. The full quote continues: "And this for some moments will quiver right over the hive, with prodigious rustle of gossamer silks that countless electrified hands might be ceaselessly rending and stitching; it floats undulating, it trembles and flutters like a veil of gladness invisible fingers support in the sky, and wave to and fro, from the flowers to the blue, expecting sublime advent or departure. And at last one angle declines, another is lifted; the radiant mantle unites its four sunlit corners; and like the wonderful carpet the fairy-tale speaks of, that flits across space to obey its master's command, it steers its straight course, bending forward a little as though to hide in its folds the sacred presence of the future, towards the willow, the pear-tree, or lime whereon the queen has alighted; and round her each rhythmical wave comes to rest, as though on a nail of gold, and suspends its fabric of pearls and of luminous wings. And then there is silence once more; and, in an instant, this mighty tumult, this awful curtain apparently laden with unspeakable menace and anger, this bewildering golden hail that streamed upon every object near—all these become merely a great, inoffensive, peaceful cluster of bees, composed of thousands of little motionless groups, that patiently wait, as they hang from the branch of a tree, for the scouts to return who have gone in search of a place of shelter."

175: Buttel-Reepen on the mental state of the swarming bees: Buttel-Reepen, H. 1900. "Sind die Bienen Reflex-maschinen?" *Experimentelle Beiträge zur Biologie der Honigbiene* 20: 1–84. He even muses (on p. 72) whether it might constitute a form of play behavior.

176: Maeterlinck quote: Same source as above.

177: Quote by the "Bee Baron": see the Buttel-Reepen article, above.

178: Lindauer's observations on dancing bees in the swarm: Lindauer, M. 1955. "Schwarmbienen auf Wohnungs-suche." *Zeitschrift für Vergleichende Physiologie* 37: 263–324; the phrase "dirty dancers" is from Seeley, T. D. 2010. *Honeybee Democracy*. Princeton: Princeton University Press.

178: On ethomics: Reiser, M. 2009. "The ethomics era?" *Nature Methods* 6 (6): 413–14. DOI: 10.1038/nmeth0609-413.

179: Quote by von Frisch: Lindauer, M. 1985. "Personal recollections of Karl von Frisch." In *Experimental Behavioral Ecology*, eds. B. Hölldobler, M. Lindauer, 5–7. Stuttgart: G. Fischer.

179: Tom Seeley's work: in addition to the book *Honeybee Democracy*, the following sources are useful: Seeley, T. D. 1982. "How honeybees find a home." *Scientific American* 247:

158–68; Seeley, T. D., Levien, R. A. 1987. "A colony of mind—the beehive as thinking machine." *Sciences* 27: 38–43. DOI: 10.1002/j.2326-1951.1987.tb02955.x; Seeley, T. D., Buhrman, S. C. 1999. "Group decision making in swarms of honey bees." *Behavioral Ecology and Sociobiology* 45: 19–31. DOI: 10.1007/s002650050536; Seeley, T. D., Visscher, P. K. 2003. "Choosing a home: how the scouts in a honey bee swarm perceive the completion of their group decision making." *Behavioral Ecology and Sociobiology* 54 (5): 511–20. DOI: 10.1007/s00265-003-0664-6; Seeley, T. D., Visscher, P. K. 2004. "Quorum sensing during nest-site selection by honeybee swarms." *Behavioral Ecology and Sociobiology* 56 (6): 594–601. DOI: 10.1007/s00265-004-0814-5; Passino, K. M., Seeley, T. D., Visscher, P. K. 2008. "Swarm cognition in honey bees." *Behavioral Ecology and Sociobiology* 62 (3): 401–14. DOI: 10.1007/s00265-007-0468-1.

181: Stop signals in the swarm: Seeley, T. D., Visscher, P. K., Schlegel, T., Hogan, P. M., Franks, N. R., Marshall, J.A.R. 2012. "Stop signals provide cross inhibition in collective decision-making by honeybee swarms." *Science* 335 (6064): 108–11. DOI: 10.1126/science.1210361.

182: On human "swarm behavior": Dyer, J.R.G., Ioannou, C. C., Morrell, L. J., Croft, D. P., Couzin, I. D., Waters, D. A., Krause, J. 2008. "Consensus decision making in human crowds." *Animal Behaviour* 75: 461–70. DOI: 10.1016/j.anbehav.2007.05.010; Moffatt, M. W. 2019. *The Human Swarm: How Our Societies Arise, Thrive, and Fall*. New York: Basic Books.

**Chapter 9**

185: Cajal and Sánchez quote: Ramón y Cajal, S., Sánchez, D. 1915. *Contribución al conocimiento de los centros nerviosos de los insectos*. Madrid: Imprenta de Hijos de Nicolás Moya.

186: Cajal's cannon: Rapport, R. 2005. *Nerve Endings: The Discovery of the Synapse*. New York: W. W. Norton.

186: History of the neuron doctrine: Strausfeld, N. J. 2012. *Arthropod Brains: Evolution, Functional Elegance, and Historical Significance*. Cambridge, MA: The Belknap Press of Harvard University Press.

186: Neuron number in the bee brain: Witthöft, W. 1967. "Absolute Anzahl und Verteilung der Zellen im Hirn der Honigbiene." *Zeitschrift für Morphologie der Tiere* 61: 160–84. DOI: 10.1007/BF00298776; recent numbers for multiple species: Godfrey, R. K., Swartzlander, M., Gronenberg, W. 2021. "Allometric analysis of brain cell number in Hymenoptera suggests ant brains diverge from general trends." *Proceedings of the Royal Society B—Biological Sciences* 288 (1947). DOI: 10.1098/rspb.2021.0199; estimate of cell number in the human brain: see Herculano-Houzel, S. 2009. "The human brain in numbers: a linearly scaled-up primate brain." *Frontiers in Human Neuroscience* 3 (31). DOI: 10.3389/neuro.09.031.2009.

186: About neuron numbers not necessarily being a useful indicator of complexity: Chittka, L., Niven, J. 2009. "Are bigger brains better?" *Current Biology* 19: R995–1008. DOI: 10.1016/j.cub.2009.08.023.

189: Descriptions of structures in the bee brain by Félix Dujardin: Dujardin, F. 1850. "Mémoire sur le systeme nerveux des insectes." *Annales des Sciences Naturelles B—Zoologie* 14: 195–206.

190: Absence of a link between dance communication and overall brain structure: Brockmann, A., Robinson, G. E. 2007. "Central projections of sensory systems involved in honey bee dance language communication." *Brain, Behavior and Evolution* 70 (2): 125–36. DOI: 10.1159/000102974.

191: Frederick Kenyon's pioneering exploration of the bee brain: Kenyon, F. C. 1896. "The brain of the bee—a preliminary contribution to the morphology of the nervous system of the Arthropoda." *Journal of Comparative Neurology* 6: 134–210. DOI: 10.1002/cne.910060302.

193: Kenyon's biographical details are from Strausfeld, *Arthropod Brains*.

194: Diversity of neurons in the insect visual system: Stirling, P., Laughlin, S. 2015. *Principles of Neural Design*. Cambridge, MA: MIT Press; Fischbach, K. F., Dittrich, A.P.M. 1989. "The optic lobe of *Drosophila melanogaster*: 1. Golgi analysis of wild-type structure." *Cell and Tissue Research* 258 (3): 441–75. DOI: 10.1007/BF00218858; Otsuna, H., Ito, K. 2006. "Systematic analysis of the visual projection neurons of *Drosophila melanogaster*: I. Lobula-specific pathways." *Journal of Comparative Neurology* 497 (6): 928–58. DOI: 10.1002/cne.21015.

195: Types of neurons in the bee visual system: Paulk, A. C., Phillips-Portillo, J., Dacks, A. M., Fellous, J. M., Gronenberg, W. 2008. "The processing of color, motion and stimulus timing are anatomically segregated in the bumblebee brain." *Journal of Neuroscience* 28 (25): 6319–32. DOI: 10.1523/JNEUROSCI.1196-08.2008.

195: Edge-orientation neurons in the bee visual system: Yang, E.-C., Maddess, T. 1997. "Orientation-sensitive neurons in the brain of the honey bee (*Apis mellifera*)." *Journal of Insect Physiology* 43 (4): 329–36. DOI: 10.1016/s0022-1910(96)00111-4.

197: Complex pattern discrimination with simple edge detectors: Roper, M., Fernando, C., Chittka, L. 2017. "Insect bio-inspired neural network provides new evidence on how simple feature detectors can enable complex visual generalization and stimulus location invariance in the miniature brain of honeybees." *PLOS Computational Biology* 13 (2): e1005333. DOI: 10.1371/journal.pcbi.1005333.

198: Counting with simple neural networks: Vasas, V., Chittka, L. 2019. "Insect-inspired sequential inspection strategy enables an artificial network of four neurons to estimate numerosity." *Science* 11: 85–92. DOI: 10.1016/j.isci.2018.12.009; based on a sequential inspection strategy as found in bumble bees: MaBouDi, H., Dona, H.S.G., Gatto, E., Loukola, O. J., Buckley, E., Onoufriou, P. D., Skorupski, P., Chittka, L. 2020. "Bumblebees use sequential scanning of countable items in visual patterns to solve numerosity tasks." *Integrative and Comparative Biology* 60 (4): 929–42. DOI: 10.1093/icb/icaa025.

198: Complex cognitive tasks can often be achieved with very low neuron numbers: see, e.g., Beer, R. D. 2003. "The dynamics of active categorical perception in an evolved model agent." *Adaptive Behavior* 11 (4): 209–43. DOI: 10.1177/1059712303114001;

Goldenberg, E., Garcowski, J., Beer, R. D. 2004. "May we have your attention: analysis of a selective attention task." In *From Animals to Animats 8: Proceedings of the Eighth International Conference on the Simulation of Adaptive Behavior*, eds. S. Schaal, A. Ijspeert, A. Billard, S. Vijayakumar, J. Hallam, J.-A. Meyer, 49–56. Cambridge, MA: MIT Press; Cruse, H. 2003. "A recurrent neural network for landmark based navigation." *Biological Cybernetics* 88: 425–37. DOI: 10.1007/s00422-003-0395-9; Cruse, H., Hübner, D. 2008. "Self-organizing memory: active learning of landmarks used for navigation." *Biological Cybernetics* 99: 219–36. DOI: 10.1007/s00422-008-0256-7; Dehaene, S., Changeux, J. P. 1993. "Development of elementary numerical abilities: a neuronal model." *Journal of Cognitive Neuroscience* 5: 390–407. DOI: 11.1162/jocn.1993.5.4.390; Dehaene, S., Changeux, J.-P., and Nadal, J. P. 1987. "Neural networks that learn temporal sequences by selection." *Proceedings of the National Academy of Sciences of the USA* 84 (9): 2727–31. DOI: 10.1073/pnas.84.9.2727; Vickerstaff, R. J., Di Paolo, E. A. 2005. "Evolving neural models of path integration." *Journal of Experimental Biology* 208: 3349–66. DOI: 10.1242/jeb.01772; Shanahan, M. 2006. "A cognitive architecture that combines internal simulation with a global workspace." *Consciousness and Cognition* 15: 433–49. DOI: 10.1016/j.concog.2005.11.005.

199: A single learning neuron: Hammer, M. 1993. "An identified neuron mediates the unconditioned stimulus in associative olfactory learning in honeybees." *Nature* 366, 59–63. DOI: 10.1038/366059a0.

201: Mushroom bodies as memory storage devices: Heisenberg, M. 2003. "Mushroom body memoir: From maps to models." *Nature Reviews Neuroscience* 4 (4): 266–75. DOI: 10.1038/nrn1074; Menzel, R. 2019. "Search strategies for intentionality in the honeybee brain." In *The Oxford Handbook of Invertebrate Neurobiology*, ed. J. H. Byrne, 663–84. Oxford: Oxford University Press. DOI: 10.1093/oxfordhb/9780190456757.013.27.

202: Fan-out, fan-in architecture in the mushroom bodies: Menzel, R. 2012. "The honeybee as a model for understanding the basis of cognition." *Nature Reviews Neuroscience* 13: 758–68. DOI: 10.1038/nrn3357; Szyszka, P., Ditzen, M., Galkin, A., Galizia, C. G., and Menzel, R. 2005. "Sparsening and temporal sharpening of olfactory representations in the honeybee mushroom bodies." *Journal of Neurophysiology* 94 (5): 3303–13. DOI: 10.1152/jn.00397.2005.

202: Modeled memory capacity in bee and ant brains: Peng, F., Chittka, L. 2017. "A simple computational model of the bee mushroom body can explain seemingly complex forms of olfactory learning and memory." *Current Biology* 27 (2): 224–30. DOI: 10.1016/j.cub.2016.10.054; Ardin, P., Peng, F., Mangan, M., Lagogiannis, K., Webb, B. 2016. "Using an insect mushroom body circuit to encode route memory in complex natural environments." *PLOS Computational Biology* 12 (2): e1004683. DOI: 10.1371/journal.pcbi.1004683.

204: For other relatively simple neural models with impressive behavioral output capacities, see, e.g.: Montague, P. R., Dayan, P., Person, C., Sejnowski, T. J. 1995. "Bee foraging in uncertain environments using predictive Hebbian learning." *Nature* 377 (6551): 725–28. DOI: 10.1038/377725a0; Shlizerman, E., Phillips-Portillo, J., Forger, D. B., Reppert, S. M. 2016. "Neural integration underlying a time-compensated sun compass in the migratory monarch butterfly." *Cell Reports* 15 (4): 683–91. DOI: 10.1016/j.celrep.2016.03.057.

205: The architecture and function of the insect central complex: Honkanen, A., Adden, A., Freitas, J. D., Heinze, S. 2019. "The insect central complex and the neural basis of navigational strategies." *Journal of Experimental Biology* 222. DOI: 10.1242/jeb.188854; Homberg, U., Heinze, S., Pfeiffer, K., Kinoshita, M., El Jundi, B. 2011. "Central neural coding of sky polarization in insects." *Philosophical Transactions of the Royal Society B—Biological Sciences* 366 (1565): 680–87. DOI: 10.1098/rstb.2010.0199; Heinze, S., Homberg, U. 2007. "Maplike representation of celestial E-vector orientations in the brain of an insect." *Science* 315 (5814): 995–97. DOI: 10.1126/science.1135531; Turner-Evans, D. B., Jayaraman, V. 2016. "The insect central complex." *Current Biology* 26 (11): R453–57. DOI: 10.1016/j.cub.2016.04.006; Gkanias, E., Risse, B., Mangan, M., Webb, B. 2019. "From skylight input to behavioural output: a computational model of the insect polarised light compass." *PLOS Computational Biology* 15 (7). DOI: 10.1371/journal.pcbi.1007123; Fisher, Y. E., Lu, F. J., D'Alessandro, I., Wilson, R. I. 2019. "Sensorimotor experience remaps visual input to a heading-direction network." *Nature* 576 (7785): 121–25. DOI: 10.1038/s41586-019-1772-4; Stone, T., Webb, B., Adden, A., Ben Weddig, N., Honkanen, A., Templin, R., Wcislo, W., Scimeca, L., Warrant, E., Heinze, S. 2017. "An anatomically constrained model for path integration in the bee brain." *Current Biology* 27 (20): 3069–85. DOI: 10.1016/j.cub.2017.08.052.

207: Central complex and consciousness: Barron, A. B., Klein, C. 2016. "What insects can tell us about the origins of consciousness." *Proceedings of the National Academy of Sciences of the USA* 113 (18): 4900–4908. DOI: 10.1073/pnas.1520084113.

208: Jewel wasps and cockroaches: Arvidson, R., Kaiser, M., Lee, S. S., Urenda, J. P., Dai, C., Mohammed, H., Nolan, C., Pan, S. Q., Stajich, J. E., Libersat, F., Adams, M. E. 2019. "Parasitoid jewel wasp mounts multipronged neurochemical attack to hijack a host brain." *Molecular & Cellular Proteomics* 18 (1): 99–114. DOI: 10.1074/mcp.RA118.000908; Hughes, D. P., Libersat, F. 2019. "Parasite manipulation of host behavior." *Current Biology* 29 (2): R45–47. DOI: 10.1016/j.cub.2018.12.001.

208: Basic consciousness-like phenomena modeled with low neuron numbers: Shanahan, "A cognitive architecture that combines internal simulation with a global workspace."

209: Bees in virtual reality: Paulk, A. C., Stacey, J. A., Pearson, T.W.J., Taylor, G. J., Moore, R.J.D., Srinivasan, M. V., van Swinderen, B. 2014. "Selective attention in the honeybee optic lobes precedes behavioral choices." *Proceedings of the National Academy of*

*Sciences of the USA* 111 (13): 5006–11. DOI: 10.1073/pnas.1323297111.

209: Neural oscillations in the bee brain: Yap, M.H.W., Grabowska, M. J., Rohrscheib, C., Jeans, R., Troup, M., Paulk, A. C., van Alphen, B., Shaw, P. J., van Swinderen, B. 2017. "Oscillatory brain activity in spontaneous and induced sleep stages in flies." *Nature Communications* 8: 1815. DOI: 10.1038/s41467-017-02024-y; Schuppe, H. 1995. "Rhythmic brain activity in sleeping bees." *Wiener Medizinische Wochenschrift* 145: 463–64.

209: Synchronizing brain areas' activity: Engel, A. K., Fries, P. 2010. "Beta-band oscillations—signalling the status quo?" *Current Opinion in Neurobiology* 20 (2): 156–65. DOI: 10.1016/j.conb.2010.02.015.

210: The exploration of sleep in bees: Kaiser, W. 1988. "Busy bees need rest, too—behavioural and electromyographical sleep signs in honeybees." *Journal of Comparative Physiology A* 163: 565–84. DOI: 10.1007/BF00603841; Kaiser, W., Steiner-Kaiser, J. 1988. "Behavioral and physiological changes occurring during sleep in the honey bee." In *Sleep 1986*, eds. W. Koella, F. Obál, H. Schulz, P. Visser, 157–59. Stuttgart: Fischer; Eban-Rothschild, A. D., Bloch, G. 2008. "Differences in the sleep architecture of forager and young honeybees (*Apis mellifera*). *Journal of Experimental Biology* 211 (15): 2408–16. DOI: 10.1242/jeb.016915.

210: Sleep and memory consolidation in bees: Klein, B. A., Klein, A., Wray, M. K., Mueller, U. G., Seeley, T. D. 2010. "Sleep deprivation impairs precision of waggle dance signaling in honey bees." *Proceedings of the National Academy of Sciences of the USA* 107 (52): 22705–9. DOI: 10.1073/pnas.1009439108; Zwaka, H., Bartels, R., Gora, J., Franck, V., Culo, A., Gotsch, M., Menzel, R. 2015. "Context odor presentation during sleep enhances memory in honeybees." *Current Biology* 25 (21): 2869–74. DOI: 10.1016/j.cub.2015.09.069.

210: For Elizabeth Tibbetts and her team's work on face recognition in *Polistes* wasps, see, e.g.: Sheehan, M. J., Tibbetts, E. A. 2008. "Robust long-term social memories in a paper wasp." *Current Biology* 18 (18): R851–52. DOI: 10.1016/j.cub.2008.07.032; Sheehan, M. J., Tibbetts, E. A. 2011. "Specialized face learning is associated with individual recognition in paper wasps." *Science* 334 (6060): 1272–75. DOI: 10.1126/science.1211334; for a popular scientific account of the latter study, see: Chittka, L., Dyer, A. 2012. "Your face looks familiar." *Nature* 481: 154–55. DOI: 10.1038/481154a; Tibbetts, E. A., Pardo-Sanchez, J., Ramirez-Matias, J., Avargues-Weber, A. 2021. "Individual recognition is associated with holistic face processing in *Polistes* paper wasps in a species-specific way." *Proceedings of the Royal Society B—Biological Sciences* 288 (1943). DOI: 10.1098/rspb.2020.3010; Tibbetts, E. A., Wong, E., Bonello, S. 2020. "Wasps use social eavesdropping to learn about individual rivals." *Current Biology* 30 (15): 3007–10.e2. DOI: 10.1016/j.cub.2020.05.053.

211: Transitive inference in paper wasps: Tibbetts, E. A., Agudelo, J., Pandit, S., Riojas, J. 2019. "Transitive inference in *Polistes* paper wasps." *Biology Letters* 15 (5). DOI: 10.1098/rsbl.2019.0015.

211: No differences in gross neuroanatomy between face-recognizing and non-face-recognizing wasps: Gronenberg, W., Ash, L. E., Tibbetts, E. A. 2008. "Correlation between facial pattern recognition and brain composition in paper wasps." *Brain, Behavior and Evolution* 71 (1): 1–14. DOI: 10.1159/000108607; however, the anterior optic tubercle, a relay for visual information processing in the central brain, grows differentially in size in individual wasps that had social exposure early in life, but not in socially isolated individuals: Jernigan, C. M., Zaba, N. C., Sheehan, M. J. 2021. "Age and social experience induced plasticity across brain regions of the paper wasp *Polistes fuscatus*." *Biology Letters* 17 (4). DOI: 10.1098/rsbl.2021.0073.

211: The human brain, compared to other primates' brains: Herculano-Houzel, S. 2012. "The remarkable, yet not extraordinary, human brain as a scaled-up primate brain and its associated cost." *Proceedings of the National Academy of Sciences of the USA* 109: 10661–68. DOI: 10.1073/pnas.1201895109.

211: Small changes in neural circuitry leading to profound changes in behavior: Katz, P. S. 2011. "Neural mechanisms underlying the evolvability of behaviour." *Philosophical Transactions of the Royal Society B* 366: 2086–99. DOI: 10.1098/rstb.2010.0336; Chittka, L., Rossiter, S. J., Skorupski, P., Fernando, C. 2012. "What is comparable in comparative cognition?" *Philosophical Transactions of the Royal Society B—Biological Sciences* 367 (1603): 2677–85. DOI: 10.1098/rstb.2012.0215 (and references therein).

**Chapter 10**

213: Quotation from Christy, R. M. 1884. "On the methodic habits of insects when visiting flowers." *Journal of the Linnean Society* 17: 186–94.

214: For examples of Charles Turner's work on individual differences: Turner, C. H. 1907. "The homing of ants: an experimental study of ant behavior." *Journal of Comparative Neurology and Psychology* 17: 367–434. DOI: 10.1002/cne.920170502; Turner, C. H. 1913. "Behavior of the common roach (*Periplaneta orientalis* L.) on an open maze." *Biological Bulletin* 25: 380–97.

214: Variation in individual behavior and intelligence—see, e.g., the following reviews: Thomson, J. D., Chittka, L. 2001. "Pollinator individuality: when does it matter?" In *Cognitive Ecology of Pollination*, eds. L. Chittka, J. D. Thomson, 191–213. Cambridge: Cambridge University Press; Jandt, J. M., Bengston, S., Pinter-Wollman, N., Pruitt, J. N., Raine, N. E., Dornhaus, A., Sih, A. 2014. "Behavioural syndromes and social insects: personality at multiple levels." *Biological Reviews* 89 (1): 48–67. DOI: 10.1111/brv.12042.

214: Variation in behavior and efficiency of division of labor: e.g., Mattila, H. R., Seeley, T. D. 2007. "Genetic diversity in honey bee colonies enhances productivity and fitness." *Science* 317: 362–64. DOI: 10.1126/science.1143046; Chittka, L., Muller, H. 2009. "Learning, specialization, efficiency and task allocation in social insects." *Communicative & Integrative Biology* 2: 151–54. DOI: 10.4161/cib.7600; Burns, J. G., Dyer, A. G. 2008. "Diversity of speed-accuracy strategies benefits social insects." *Current Biology* 18: R953–54. DOI: 10.1016/j.cub.2008.08.028; Muller, H., Chittka, L. 2008. "Animal personalities: the advantage of diversity." *Current Biology* 20: R961–63. DOI: 10.1016/jcub.2008.09.001; Cook, C. N., Lemanski, N. J., Mosqueiro, T., Ozturk, C., Gadau, J., Pinter-Wollman, N., Smith, B. H. 2020. "Individual learning phenotypes drive collective behavior." *Proceedings of the National Academy of Sciences of the USA* 117 (30): 17949–56. DOI: 10.1073/pnas.1920554117.

217: Bumble bee foraging under permanent daylight conditions: Stelzer, R. J., Chittka, L. 2010. "Bumblebee foraging rhythms under the midnight sun measured with radiofrequency identification." *BMC Biology* 8: 93. DOI: 10.1186/1741-7007-8-93.

217: Microchipped bumble bees and individual differences: Stelzer, R. J., Stanewsky, R., Chittka, L. 2010. "Circadian foraging rhythms of bumblebees monitored by radiofrequency identification." *Journal of Biological Rhythms* 25: 257–67. DOI: 10.1177/0748730410371750.

217: "Death dance" in bumble bees—see Stelzer et al. reference above—and fruit flies: Tower, J., Agrawal, S., Alagappan, M. P., Bell, H. S., Demeter, M., Havanoor, N., Hegde, V. S., Jia, Y. D., Kothawade, S., Lin, X. Y., et al. 2019. "Behavioral and molecular markers of death in *Drosophila melanogaster*." *Experimental Gerontology* 126. DOI: 10.1016/j.exger.2019.110707.

220: Physiological and behavioral differences between bee queens and workers: Chittka, A., Chittka, L. 2010. "Epigenetics of royalty." *PLOS Biology* 8 (11). DOI: 10.1371/journal.pbio.1000532 (and references therein).

220: Differences in mushroom body volume over a honey bee's lifetime: Durst, C., Eichmüller, S., Menzel, R. 1994. "Development and experience lead to increased volume of subcompartments of the honeybee mushroom body." *Behavioral and Neural Biology* 62: 259–63. DOI: 10.1016/S0163-1047(05)80025-1; Fahrbach, S. E., Moore, D., Capaldi, E. A., Farris, S. M., Robinson, G. E. 1998. "Experience-expectant plasticity in the mushroom bodies of the honeybee." *Learning & Memory* 5: 115–23.

221: François Huber on self-organization and climate control: Huber, F. 1814. *Nouvelles observations sur les abeilles* (2nd edition); trans. C. P. Dadant, as *New Observations upon Bees*. 1926. Hamilton, IL: *American Bee Journal*.

222: Individual sensitivity and task specialization in an insect colony: Beshers, S. N., Fewell, J. H. 2001. "Models of division of labor in social insects." *Annual Review of Entomology* 46: 413–40. DOI: 10.1146/annurev.ento.46.1.413; Jeanson, R., Clark, R. M., Holbrook, C. T., Bertram, S. M., Fewell, J. H., Kukuk, P. F. 2008. "Division of labour and socially induced changes in response thresholds in associations of solitary halictine bees." *Animal Behaviour* 7 (3): 593–602. DOI: 10.1016/j.anbehav.2008.04.007; Page, R. E., Robinson, G. E., Fondrk, M. K. 1989. "Genetic specialists, kin recognition and nepotism in honey-bee colonies." *Nature* 338: 576–79. DOI: 10.1038/338576a0; but see also: Ulrich, Y., Kawakatsu, M., Tokita, C. K.,

Saragosti, J., Chandra, V., Tarnita, C. E., Kronauer, D.J.C. 2021. "Response thresholds alone cannot explain empirical patterns of division of labor in social insects." *PLOS Biology* 19 (6). DOI: 10.1371/journal.pbio.3001269.

222: Jennifer Fewell on "dishwashing experts": Fewell, J. H. 2003. "Social insect networks." *Science* 301 (5641): 1867–70. DOI: 10.1126/science.1088945.

223: Karl von Frisch's work about the sense of taste in honey bees: von Frisch, K. 1934. "Über den Geschmackssinn der Bienen." *Zeitschrift für Vergleichende Physiologie* 21: 1–156. DOI: 10.1007/BF00338271.

223: Robert Page and team on bees' sugar sensitivity and individuality: Page, R. E., Schneir, R., Erber, J., Amdam, G. V. 2006. "The development and evolution of division of labor and foraging specialization in a social insect (*Apis mellifera* L.)." *Current Topics in Developmental Biology* 74: 253–86. DOI: 10.1016/S0070-2153(06)74008-X.

224: Size-dependent labor division: Spaethe, J., Weidenmüller, A. 2002. "Size variation and foraging rate in bumblebees (*Bombus terrestris*)." *Insectes Sociaux* 49: 142–46. DOI: 10.1007/s00040-002-8293-z; a parallel study from Dave Goulson's team: Goulson, D., Peat, J., Stout, J. C., Tucker, J., Darvill, B., Derwent, L. C., Hughes, W.O.H. 2002. "Can alloethism in workers of the bumblebee, *Bombus terrestris*, be explained in terms of foraging efficiency?" *Animal Behaviour* 64: 123–30. DOI: 10.1006/anbe.2002.3041.

224: Larger individuals have better vision: Spaethe, J., Chittka, L. 2003. "Interindividual variation of eye optics and single object resolution in bumblebees." *Journal of Experimental Biology* 206 (19): 3447–53. DOI: 10.1242/jeb.00570.

225: Large individuals have higher odor sensitivity: Spaethe, J., Brockmann, A., Halbig, C., Tautz, J. 2007. "Size determines antennal sensitivity and behavioral threshold to odors in bumblebee workers." *Naturwissenschaften* 94: 733–39. DOI: 10.1007/s00114-007-0251-1.

226: Task specialization as a result of experience: Ravary, F., Lecoutey. E., Kaminski, G., Chaline, N., Jaisson, P. 2007. "Individual experience alone can generate lasting division of labor in ants." *Current Biology* 17 (15): 1308–12. DOI: 10.1016/j.cub.2007.06.047.

227: Individually different foraging routes: Heinrich, B. 1976. "The foraging specializations of individual bumblebees." *Ecological Monographs* 46 (2): 105–28. DOI: 10.2307/1942246; Thomson, J. D., Maddison, W. P., Plowright, R. C. 1982. "Behavior of bumble bee pollinators on *Aralia hispida* Vent. (Araliaceae)." *Oecologia* 54: 326–36. DOI: 10.1007/BF00380001; Thomson and Chittka, "Pollinator individuality: when does it matter?"; Saleh, N., Chittka, L. 2007. "Traplining in bumblebees (*Bombus impatiens*): a foraging strategy's ontogeny and the importance of spatial reference memory in short-range foraging." *Oecologia* 151 (4): 719–30. DOI: 10.1007/s00442-006-0607-9; Lihoreau, L., Chittka, L., Raine, N. E. 2010. "Travel optimization by foraging bumblebees through readjustments of traplines after discovery of new feeding locations." *American Naturalist* 176 (6): 744–57. DOI: 10.1086/657042; Lihoreau, M. D., Raine, N. E., Reynolds, A. M., Stelzer, R. J., Lim, K. S., Smith, A. D., Osborne, J. L., Chittka, L. 2012. "Radar tracking and motion-sensitive cameras on flowers reveal the development of pollinator multi-destination routes over large spatial scales." *PLOS Biology* 10: e1001392. DOI: 10/1371/journal.pbio.1001392.

229: Charles Turner's suggestion of speed-accuracy tradeoffs: Turner, C. H. 1913. "Behavior of the common roach (*Periplaneta orientalis* L.) on an open maze." *Biological Bulletin* 25: 348–65.

230: Speed-accuracy tradeoff differences between individual bees: Chittka, L., Dyer, A. G., Bock, F., Dornhaus, A. 2003. "Bees trade off foraging speed for accuracy." *Nature* 424: 388. DOI: 10.1038/424388a; Wang, M., Chittka, L., Ings, T. C. 2018. "Bumblebees express consistent, but flexible, speed-accuracy tactics under different levels of predation threat." *Frontiers in Psychology* 9: 1601. DOI: 10.3389/fpsyg.2018.01601.

230: On how individual variation may benefit the colony as a whole: Burns, J. G., Dyer, A. G. 2008. "Diversity of speed-accuracy strategies benefits social insects." *Current Biology* 18 (20): R953–54. DOI: 10.1016/j.cub.2008.08.028; Mattila, H. R., Seeley, T. D. 2007. "Genetic diversity in honey bee colonies enhances productivity and fitness." *Science* 317: 362–64. DOI: 10.1126/science.1143046; Chittka, L., Skorupski, P., Raine, N. E. 2009. "Speed-accuracy tradeoffs in animal decision making." *Trends in Ecology & Evolution* 24 (7): 400–407. DOI: 10.1016/j.tree.2009.02.010.

231: Study in which we observed a bee that voluntarily flew into a black container: Raine, N. E., Chittka, L. 2008. "The correlation of learning speed and natural foraging success in bumble-bees." *Proceedings of the Royal Society B—Biological Sciences* 275: 803–8. DOI: 10.1098/rspb.2007.1652.

231: Behavioral variability and problem solving: Brembs, B. 2011. "Towards a scientific concept of free will as a biological trait: spontaneous actions and decision-making in invertebrates." *Proceedings of the Royal Society B—Biological Sciences* 278: 930–39. DOI: 10.1098/rspb.2010.2325.

232: Individual variation in the string-pulling study: Alem, S., Perry, C. J., Zhu, X., Loukola, O. J., Ingraham, T., Søvik, E., Chittka, L. 2016. "Associative mechanisms allow for social learning and cultural transmission of string pulling in an insect." *PLOS Biology* 14 (10). DOI: 10.1371/journal.pbio.1002564.

232: Graded differences in learning speed between individuals and individual learning curves were first shown for insects by Charles Turner, who followed the same individual ants through multiple trials at a homing task. Unfortunately, this work was ignored for decades, so that well into the 1990s people often reported learning curves for insects by reporting changes in group behavior—a substantial complication, since learning by definition only happens in individuals. Turner reference: Turner, C. H. 1907. "The homing of ants: an experimental study of ant behavior."

*Journal of Comparative Neurology and Psychology* 17: 367–434. DOI: 10.1002/cne.920170502.

232: Using mathematical tools for quantifying learning performance: Chittka, L., Thomson, J. D. 1997. "Sensori-motor learning and its relevance for task specialization in bumble bees." *Behavioral Ecology and Sociobiology* 41: 385–98. DOI: 10.1007/s00265 0050400; Raine, N. E., Chittka, L. 2008. "The correlation of learning speed and natural foraging success in bumble-bees." *Proceedings of the Royal Society B—Biological Sciences* 275: 803–8.

233: Correlation of learning performance across different tasks: Muller, H., Chittka, L. 2012. "Consistent interindividual differences in discrimination performance by bumblebees in colour, shape and odour learning tasks (Hymenoptera: Apidae: *Bombus terrestris*)." *Entomologia Generalis* 34: 1–8.

233: On general intelligence: Burkart, J. M., Schubiger, M. N., van Schaik, C. P. 2017. "The evolution of general intelligence." *Behavioral and Brain Sciences* 40: E195. DOI: 10.1017 /s0140525x16000959.

233: Biographical details of Oskar Vogt are from: Klatzo, I. 2002. *Cécile and Oskar Vogt: The Visionaries of Modern Neuroscience*. Berlin: Springer Verlag. The Vogts' biography highlights the train of thought from the early bumble bee observations to the later study of human brain variation: "Since . . . genetic studies, at that time, showed that in insects changes in external markings (e.g. hair patterns in bumble-bees) can be due to . . . *environmental factors*, the Vogts became inclined to think that the variability in ontogenetic formations of various human brain regions could also be of a similar origin."

234: Conceptual links between variation in animal and human brains: Vogt, C., Vogt, O. 1937. "Sitz und Wesen der Krankheiten im Lichte der topistischen Hirnforschung und des Variieres der Tiere. Erster Teil. Befunde der topistischen Hirnforschung als Beitrag zur Lehre vom Krankheitssitz." *Journal für Psychologie und Neurologie* 47: 237–457; Vogt, C., Vogt, O. 1938. "Sitz und Wesen der Krankheiten . . . Zweiter Teil, 1. Hälfte. Zur Einführung in das Variieren der Tiere. Die Erscheinungsseiten der Variation." *Journal für Psychologie und Neurologie* 48: 169–324.

234: Oskar Vogt's work on variation in bumble bees, and bumble bees as a model for understanding evolutionary processes: Vogt, O. 1911. "Studien über das Artproblem. Über das Variieren der Hummeln. 2. Teil. *Sitzungsberichte der Gesellschaft Naturforschender Freunde zu Berlin* 1911: 31–74.

237: Difference in bumble bee brain structure and learning/memory performance: Li, L., MaBouDi, H., Egertova, M., Elphick, M. R., Chittka, L., Perry, C. J. 2017. "A possible structural correlate of learning performance on a colour discrimination task in the brain of the bumblebee." *Proceedings of the Royal Society B—Biological Sciences* 284 (1864). DOI: 10.1098 /rspb.2017.1323.

238: Selection experiments on learning performance by honey bees and earlier work in flies: Brandes, C., Frisch, B., Menzel, R. 1988. "Time-course of memory formation differs in honey bee lines selected for good and poor learning." *Animal Behaviour* 36: 981–85; McGuire, T. R., Hirsch, J. 1977. "Behavior-genetic analysis of *Phormia regina*: conditioning, reliable individual differences, and selection." *Proceedings of the National Academy of Sciences of the USA* 74: 5193–97; Lofdahl, K. L., Holliday, M., Hirsch, J. 1992. "Selection for conditionability in *Drosophila melanogaster*." *Journal of Comparative Psychology* 106: 172–83.

239: Learning curves for each individual, and colony performance under field conditions: Raine, N. E., Chittka, L. 2008. "The correlation of learning speed and natural foraging success in bumble-bees." *Proceedings of the Royal Society B—Biological Sciences* 275: 803–8. DOI: 10.1098/rspb.2007.1652.

240: Smart learners doing well at all tasks: Muller, H., Chittka, L. 2012. "Consistent interindividual differences in discrimination performance by bumblebees in colour, shape and odour learning tasks (Hymenoptera: Apidae: *Bombus terrestris*)." *Entomologia Generalis* 34: 1–8; Raine, N. E., Chittka, L. 2012. "No trade-off between learning speed and associative flexibility in bumblebees: a reversal learning test with multiple colonies." *PLOS One* 7: e45096. DOI: 10.1371/journal.pone.0045096.

240: Energetic cost of learning: Dukas, R. 1999. "Costs of memory: ideas and predictions." *Journal of Theoretical Biology* 197: 41–50. DOI: 10.1006/jtbi.1998.0856; Snell-Rood, E. C., Papaj, D. R., Gronenberg, W. 2009. "Brain size: a global or induced cost of learning?" *Brain, Behavior and Evolution* 73 (2): 111–28. DOI: 10.1159/000213647; Evans, L. J., Smith, K. E., Raine, N. E. 2017. "Fast learning in free-foraging bumble bees is negatively correlated with lifetime resource collection." *Scientific Reports* 7: 496. DOI: 10.1038/s41598-017-00389-0; but see also: Liefting, M., Rohmann, J. L., Le Lann, C., Ellers, J. 2019. "What are the costs of learning? Modest trade-offs and constitutive costs do not set the price of fast associative learning ability in a parasitoid wasp." *Animal Cognition* 22: 851–61. DOI: 10.1007/s10071-019-01281-2.

**Chapter 11**

242: Bonnet quotation: Bonnet, C. 1764. *Contemplation de la nature*. Amsterdam: Marc Michel Rey (part XI, notes 9 and 11 of the XXVII chapter, last edition), cited in Huber, F. 1814. *Nouvelles observations sur les abeilles* (2nd edition); trans. C. P. Dadant, as *New Observations upon Bees*. 1926. Hamilton, IL: *American Bee Journal* (p. 105). Curiously, Bonnet writes these words to explain how the honey bee comb comes about by simple processes, and denounces the bees as automatons—but at the same time furnishes them with consciousness and feelings.

244: Karl von Frisch on the absence of pain in honey bees: "A bee sits at the feeder and imbibes sugar water. You cut off her abdomen at the thin waistline with scissors. Her head and thorax stay in place and the meal proceeds, only that . . . everything leaks out at the back. A little lake of sugar water grows in that place where the abdomen belongs . . . because the bee never satiates and keeps sucking, until she keels over, exhausted, but ending her life in pleasure. Such behavior is incompatible with the perception of pain. This would simply not make sense in animals with a hard exoskeleton. In us, with our soft skin, pain is a life-saving warning sign that ensures that we duly avoid injury." Translated by the author from: von Frisch, K. 1959. "Insekten—die Herren der Erde." *Naturwissenschaftliche Rundschau* 10: 369–75.

244: Segregated pathways for nociception and mechanoreception: Burrell, B. D. 2017. "Comparative biology of pain: what invertebrates can tell us about how nociception works." *Journal of Neurophysiology* 117 (4): 1461–73. DOI: 10.1152/jn.00600.2016.

245: Nociception in bees and other invertebrates: Tobin, D. M., Bargmann, C. I. 2004. "Invertebrate nociception: behaviors, neurons and molecules." *Journal of Neurobiology* 61 (1): 161–74. DOI: 10.1002/neu.20082; Junca, P., Sandoz, J.-C. 2015. "Heat perception and aversive learning in honey bees: putative involvement of the thermal / chemical sensor AmHsTRPA. *Frontiers in Physiology* 6: 316. DOI: 10.3389/fphys.2015.00316.

246: Wound healing in insects: Frank, E. T., Wehrhahn, M., Linsenmair, K. E. 2018. "Wound treatment and selective help in a termite-hunting ant." *Proceedings of the Royal Society B—Biological Sciences* 285 (1872). DOI: 10.1098/rspb.2017.2457; Frank, E. T., Schmitt, T., Hovestadt, T., Mitesser, O., Stiegler, J., Linsenmair, K. E. 2017. "Saving the injured: rescue behavior in the termite-hunting ant *Megaponera analis*." *Science Advances* 3 (4). DOI: 10.1126/sciadv.1602187.

246: Distinguishing nociception and pain: Elwood, R. W. 2019. "Discrimination between nociceptive reflexes and more complex responses consistent with pain in crustaceans." *Philosophical Transactions of the Royal Society B—Biological Sciences* 374 (1785). DOI: 10.1098/rstb.2019.0368.

247: On the impossibility of measuring pain or suffering objectively: Mendl, M., Paul, E. S., Chittka, L. 2011. "Animal behaviour: emotion in invertebrates?" *Current Biology* 21 (12): R463–65. DOI: 10.1016/j.cub.2011.05.028 (and references therein).

248: Experiments on the effects of alarm pheromone on responses to aversive stimuli: Nuñez, J., Almeida, L., Balderrama, N., Giurfa, M. 1997. "Alarm pheromone induces stress analgesia via an opioid system in the honeybee." *Physiology & Behavior* 63 (1): 75–80. DOI: 10.1016/s0031-9384(97)00391-0.

249: Absence of an endogenous opiate system in insects: Elphick, M. R., Mirabeau, O., Larhammar, D. 2018. "Evolution of neuropeptide signalling systems." *Journal of Experimental Biology* 221 (3). DOI: 10.1242/jeb.151092.

249: Possibility of opiates binding to non-opiate receptors in insects: Koyyada, R., Latchooman, N., Jonaitis, J., Ayoub, S. S., Corcoran, O., Casalotti, S. O. 2018. "Naltrexone reverses ethanol preference and protein kinase C activation in *Drosophila melanogaster*". *Frontiers in Physiology* 9: 175. DOI: 10.3389/fphys.2018.00175.

249: Allatostatins in honey bees: Urlacher, E., Devaud, J. M., Mercer, A. R. 2019. "Changes in responsiveness to allatostatin treatment accompany shifts in stress reactivity in young worker honey bees." *Journal of Comparative Physiology A—Neuroethology, Sensory, Neural, and Behavioral Physiology* 205 (1): 51–59. DOI: 10.1007/s00359-018-1302-0; Urlacher, E., Devaud, J. M., Mercer, A. R. 2017. "C-type allatostatins mimic stress-related effects of alarm pheromone on honey bee learning and memory recall." *PLOS One* 12 (3). DOI: 10.1371/journal.pone.0174321; Urlacher, E., Soustelle, L., Parmentier, M. L., Verlinden, H., Gherardi, M. J., Fourmy, D., Mercer, A. R., Devaud, J. M., Massou, I. 2016. "Honey bee allatostatins target galanin / somatostatin-like receptors and modulate learning: a conserved function? *PLOS One* 11 (1). DOI: 10.1371/journal.pone.0146248.

250: Psychological responses of bees to predation attacks from crab spiders: Ings, T. C., Chittka, L. 2008. "Speed-accuracy tradeoffs and false alarms in bee responses to cryptic predators." *Current Biology* 18: 1520–24. DOI: 10.1016/j.cub.2008.07.074; see also: Jones, E. I., Dornhaus, A. 2011. "Predation risk makes bees reject rewarding flowers and reduce foraging activity." *Behavioral Ecology and Sociobiology* 65 (8): 1505–11. DOI: 10.1007/s00265-011-1160-z; Huey, S., Nieh, J. C. 2017. "Foraging at a safe distance: crab spider effects on pollinators." *Ecological Entomology* 42 (4): 469–76. DOI: 10.1111/een.12406.

253: Melissa Bateson and Geraldine Wright's work on cognitive bias in honey bees: Bateson, M., Desire, S., Gartside, S. E., Wright, G. A. 2011. "Agitated honeybees exhibit pessimistic cognitive biases." *Current Biology* 21 (12): 1070–73. DOI: 10.1016/j.cub.2011.05.017.

253: The effect of surprise rewards on emotion-like states: Solvi, C., Baciadonna, L., Chittka, L. 2016. "Unexpected rewards induce dopamine-dependent positive emotion-like state changes in bumblebees." *Science* 353 (6307): 1529–31. DOI: 10.1126/science.aaf4454.

255: Insects seeking out mood-altering substances: Chittka, L., Wilson, C. 2019. "Expanding consciousness." *American Scientist* 107 (6): 364–69. DOI: 10.1511/2019.107.6.364 (and references therein).

255: Male fruit flies' rewarding ejaculations and alcohol-seeking: Shir Zer-Krispil, S., Zak, H., Shao, L., Bentzur, A., Shmueli, A., Shohat-Ophir, G. 2018. "Ejaculation induced by the activation of Crz neurons is rewarding to *Drosophila* males." *Current Biology* 28 (9): 1445–52. DOI: 10.1016/j.cub.2018.03.039; Shohat-Ophir, G., Kaun, K. R., Azanchi, R., Heberlein, U. 2012. "Sexual deprivation increases ethanol intake in *Drosophila*." *Science* 335 (6074): 1351–55. DOI: 10.1126/science.1215932.

255: Bees preferring nectar with caffeine or nicotine: Wright, G. A., Baker, D. D., Palmer, M. J., Stabler, D., Mustard, J. A., Power, E. F., Borland, A. M., Stevenson, P. C. 2013. "Caffeine in floral nectar enhances a pollinator's memory of reward." *Science* 339 (6124): 1202–4. DOI: 10.1126/science.1228806. For a popular scientific account, see: Chittka, L., Peng, F. 2013. "Caffeine boosts bees' memories." *Science* 339 (6124): 1157–59. DOI: 10.1126/science.1234411; Baracchi, D.,

Marples, A., Jenkins, A. J., Leitch, A. R., Chittka, L. 2017. "Nicotine in floral nectar pharmacologically influences bumblebee learning of floral features." *Scientific Reports* 7: 1951. DOI: 10.1038/s41598-017-01980-1.

257: Philosophical thoughts about an ancient evolution of consciousness: Godfrey-Smith, P. 2017. *Other Minds: The Octopus and the Evolution of Intelligent Life*. London: William Collins; Bronfman, Z. Z., Ginsburg, S., Jablonka, E. 2016. "The evolutionary origins of consciousness suggesting a transition marker." *Journal of Consciousness Studies* 23 (9–10): 7–34.

257: Mirror self-recognition: Gallup, G. G., Povinelli, D. J., Suarez, S. D., Anderson, J. R., Lethmate, J., Menzel, E. W. 1995. "Further reflections on self-recognition in primates." *Animal Behaviour* 50: 1525–32. DOI: 10.1016/0003-3472(95)80008-5; Reiss, D., Marino, L. 2001. "Mirror self-recognition in the bottlenose dolphin: a case of cognitive convergence." *Proceedings of the National Academy of Sciences of the USA* 98 (10): 5937–42. DOI: 10.1073/pnas.101086398; Plotnik, J. M., de Waal, F.B.M., Reiss, D. 2006. "Self-recognition in an Asian elephant." *Proceedings of the National Academy of Sciences of the USA* 103 (45): 17053–57. DOI: 10.1073/pnas.0608062103.

258: Bumble bees aware of their own body dimensions: Ravi, S., Siesenop, T., Bertrand, O., Li, L., Doussot, C., Warren, W. H., Combes, S. A., Egelhaaf, M. 2020. "Bumblebees perceive the spatial layout of their environment in relation to their body size and form to minimize inflight collisions." *Proceedings of the National Academy of Sciences of the USA* 117 (49): 31494–99. DOI: 10.1073/pnas.2016872117; for a popular scientific account, see: Brebner, J., Chittka, L. 2021. "Animal cognition: the self-image of a bumblebee." *Current Biology* 31: R207–9. DOI: 10.1016/j.cub.2020.12.027.

258: Mammals' awareness of body dimensions being linked to consciousness: Dale, R., Plotnik, J. M. 2017. "Elephants know when their bodies are obstacles to success in a novel transfer task." *Scientific Reports* 7. DOI: 10.1038/srep46309; Brownell, C. A., Zerwas, S., Ramani, G. B. 2007 "'So big': The development of body self-awareness in toddlers." *Child Development* 78 (5): 1426–40. DOI: 10.1111/j.1467-8624.2007.01075.x; Warren, W. H., Whang, S. 1987. "Visual guidance of walking through apertures—body-scaled information for affordances." *Journal of Experimental Psychology—Human Perception and Performance* 13 (3): 371–83. DOI: 10.1037/0096-1523.13.3.371.

258: The possible relationship of self-recognition and inbreeding avoidance: Capodeanu-Nagler, A., Rapkin, J., Sakaluk, S. K., Hunt, J., Steiger, S. 2014. "Self-recognition in crickets via on-line processing." *Current Biology* 24 (23): R1117–18. DOI: 10.1016/j.cub.2014.10.050.

260: Recognizing biological motion by an invertebrate: De Agrò, M., Rößler, D. C., Kim, K., Shamble, P. S. 2021. "Perception of biological motion by jumping spiders." *PLOS Biology* 19 (7): e3001172. DOI: 10.1371/journal.pbio.3001172.

261: Judging whether it's safe to jump over a stream: there is evidence from some insects that, indeed, the width of a gap to be crossed is judged accurately, even taking into account the individual capacity to cross the gap: Krause, T., Spindler, L., Poeck, B., Strauss, R. 2019. "*Drosophila* acquires a long-lasting body-size memory from visual feedback." *Current Biology* 29 (11): 1833–41.e3. DOI: 10.1016/j.cub.2019.04.037; Niven, J. E., Buckingham, C. J., Lumley, S., Cuttle, M. F., Laughlin, S. B. 2010. "Visual targeting of forelimbs in ladder-walking locusts." *Current Biology* 20 (1): 86–91. DOI: 10.1016/j.cub.2009.10.079; Niven, J. E., Ott, S. R., Rogers, S. M. 2012. "Visually targeted reaching in horse-head grasshoppers." *Proceedings of the Royal Society B—Biological Sciences* 279 (1743): 3697–705. DOI: 10.1098/rspb.2012.0918.

262: Mark Roper's work on responding appropriately to shapes without storing shapes in memory: Roper, M., Fernando, C., Chittka, L. 2017. "Insect bio-inspired neural network provides new evidence on how simple feature detectors can enable complex visual generalization and stimulus location invariance in the miniature brain of honeybees." *PLOS Computational Biology* 13 (2): e1005333. DOI: 10.1371/journal.pcbi.1005333.

262: Visual stimulus recognition without awareness in humans: Lau, H. C., Passingham, R. E. 2006. "Relative blindsight in normal observers and the neural correlate of visual consciousness." *Proceedings of the National Academy of Sciences of the USA* 103 (49): 18763–68. DOI: 10.1073/pnas.0607716103; Schmid, M. C., Mrowka, S. W., Turchi, J., Saunders, R. C., Wilke, M., Peters, A. J., Ye, F. Q., Leopold, D. A. 2010. "Blindsight depends on the lateral geniculate nucleus." *Nature* 466 (7304): 373–77. DOI: 10.1038/nature09179.

263: On Molyneux's problem: Degenaar, M., Lokhorst, G. J. 2017. "Molyneux's problem." In *The Stanford Encyclopedia of Philosophy*, ed. E. N. Zalta. Stanford, CA: Metaphysics Research Lab, Philosophy Department, Stanford University.

264: Bumble bees' cross-modal object recognition in darkness and light: Solvi, C., Gutierrez Al-Khudhairy, S., Chittka, L. 2020. "Bumble bees display cross-modal object recognition between visual and tactile senses." *Science* 367 (6480): 910–12. DOI: 10.1126/science.aay8064; a similar experiment by scientists at the University of Bristol on cross-modal object recognition also indicates that bees might indeed be able to picture the spatial arrangements of features in a pattern. Bees were first trained to distinguish two types of artificial flowers that were visually identical, but which had "invisible patterns" made up of small, scented holes that were arranged either in a circle or in a cross, and bees were able to figure out these patterns by using their feelers. If these patterns were suddenly made visible by the experimenter (so that the flowers now displayed visual circles or crosses, but no longer had any scent), bees instantly recognized the image that formerly was just an ephemeral pattern of volatiles in the air. This indicates that they had a mental representation of the shapes learned; see Lawson, D. A., Chittka, L., Whitney, H. M., Rands, S. A. 2018. "Bumblebees distinguish floral scent patterns, and can transfer these to corresponding visual patterns." *Proceedings of the Royal Society B—Biological Sciences* 285 (1880). DOI: 10.1098/rspb.2018.0661.

266: Metacognition in honey bees? See Perry, C. J., Barron, A. B. 2013. "Honey bees selectively avoid difficult choices." *Proceedings of the National Academy of Sciences of the USA* 110 (47): 1915559. DOI: 10.1073/pnas.1314571110. (Note: The first author of the study is now named Cwyn Solvi.)

267: Bees as a magic well: von Frisch, K. 1950. *Bees: Their Vision, Chemical Senses and Language*. Ithaca, NY: Cornell University Press.

267: Consciousness different in different animals: Birch, J., Schnell, A. K., Clayton, N. S. 2020. "Dimensions of animal consciousness." *Trends in Cognitive Sciences* 24 (10): 789–801. DOI: 10.1016/j.tics.2020.07.007.

**Chapter 12**

270: Bees using plastic and polystyrene as nesting materials: Prendergast, K. S. 2020. "Scientific note: mass-nesting of a native bee *Hylaeus (Euprosopoides) ruficeps kalamundae* (Cockerell, 1915) (Hymenoptera: Colletidae: Hylaeinae) in polystyrene." *Apidologie* 51 (1): 107–11. DOI: 10.1007/s13592-019-00722-8; Allasino, M. L., Marrero, H. J., Dorado, J., Torretta, J. P. 2019. "Scientific note: first global report of a bee nest built only with plastic." *Apidologie* 50 (2): 230–33. DOI: 10.1007/s13592-019-00635-6.

270: On how everyone can contribute to bee conservation, see Dave Goulson's books: e.g., Goulson, D. 2019. *The Garden Jungle: Or Gardening to Save the Planet*. New York: Vintage; Goulson, D. 2021. *Gardening for Bumblebees: A Practical Guide to Creating a Paradise for Pollinators*. London: Penguin Books; and, e.g., the Bumblebee Conservation Trust's web page: https://www.bumblebeeconservation.org/.

272: On managed honey bees interacting adversely with native pollinators in many environments: Geldmann, J., González-Varo, J. P. 2018. "Conserving honey bees does not help wildlife." *Science* 359 (6374): 392–93. DOI: 10.1126/science.aar2269; Ropars, L., Dajoz, I., Fontaine, C., Muratet, A., Geslin, B. 2019. "Wild pollinator activity negatively related to honey bee colony densities in urban context." *PLOS One* 14 (9). DOI: 10.1371/journal.pone.0222316; Fürst, M. A., McMahon, D. P., Osborne, J. L., Paxton, R. J., Brown, M.J.F. 2014. "Disease associations between honeybees and bumblebees as a threat to wild pollinators." *Nature* 506 (7488): 364–66. DOI: 10.1038/nature12977; Angelella, G. M., McCullough, C. T., O'Rourke, M. E. 2021. "Honey bee hives decrease wild bee abundance, species richness, and fruit count on farms regardless of wildflower strips." *Scientific Reports* 11 (1): 3202. DOI: 10.1038/s41598-021-81967-1; Herrera, C. M. 2020. "Gradual replacement of wild bees by honeybees in flowers of the Mediterranean Basin over the last 50 years." *Proceedings of the Royal Society B—Biological Sciences* 287 (1921). DOI: 10.1098/rspb.2019.2657.

# ILLUSTRATION CREDITS

1.1. Photo by Helga Heilmann: figure 1 in Gallo, V., Chittka, L. 2018. "Cognitive aspects of comb-building in the honeybee?" *Frontiers in Psychology* 9: 900. DOI: 10.3389/fpsyg.2018.00900.
1.2. A. Electron micrograph by Johannes Spaethe.
1.2. B and C. Images by Andrew Giger.
1.3. Photo by Lars Chittka.
1.4. Reprinted from figure 1 in *Animal Behaviour*, Vol. 36, Laverty, T. M., Plowright, R. C., "Flower handling by bumblebees—a comparison between specialists and generalists," 733–40. Copyright 1988, with permission from Elsevier.
1.5. Photo and image design by Helga Heilmann, published as the cover image to the article: Roper, M., Fernando, C., Chittka, L. 2017. "Insect bio-inspired neural network provides new evidence on how simple feature detectors can enable complex visual generalization and stimulus location invariance in the miniature brain of honeybees." *PLOS Computational Biology* 13 (2): e1005333. DOI:10.1371/journal.pcbi.1005333.
2.1. Photos by Klaus Schmitt, as figure 1b from Stelzer, R. J., Raine, N. E., Schmitt, K. D., Chittka, L. 2010. "Effects of aposematic coloration on predation risk in bumblebees? A comparison between differently coloured populations, with consideration of the ultraviolet." *Journal of Zoology* 282: 75–83. DOI: 10.1111/j.1469-7998.2010.00709.x.
2.2. Photo by Karl von Frisch; figure 1 from von Frisch, K. 1914. "Der Farbensinn und Formensinn der Biene." *Zoologische Jahrbücher (Physiologie)* 37: 1–238. DOI: 10.5962/bhl.title.11736.
2.3. Figure design by Lars Chittka.
2.4. Modified from figure 1 in Waser, N. M., Chittka, L. 1998. "Bedazzled by flowers." *Nature* 394: 835–36. DOI: 10.1038/29657.
2.5. Figure design by Lars Chittka.
3.1. Redrawn after figure 11 from Wolf, E. 1927. "Über das Heimkehrvermögen der Bienen II." *Zeitschrift für Vergleichende Physiologie* 6: 221–54. DOI: 10.1007/BF00339256.
3.2. Modified from the following sources: Wehner, R. 1997. "The ant's celestial compass system: spectral and polarization channels." In *Orientation and Communication in Arthropods*, ed. M. Lehrer, Basel: Birkhauser Verlag, 145–85; Evangelista, C., Kraft, P., Dacke, M., Labhart, T., Srinivasan, M. V. 2014. "Honeybee navigation: critically examining the role of the polarization compass." *Philosophical Transactions of the Royal Society B—Biological Sciences* 369: DOI: 10.1098/rstb.2013.0037.
3.3. Modified from Srinivasan, M. V. 2011. "Honeybees as a model for the study of visually guided flight, navigation, and biologically inspired robotics." *Physiological Reviews* 91 (2): 413–60. DOI: 10.1152/physrev.00005.2010.

3.4. Modified from figure 4A in Liang, C. H., Chuang, C. L., Jiang, J. A., Yang, E. C. 2016. "Magnetic sensing through the abdomen of the honey bee." *Scientific Reports* 6. DOI: 10.1038/srep23657.

3.5. From figure 1.1 in Goodman, L. 2003. *Form and Function in the Honeybee.* Cardiff, Westdale Press. © International Bee Research Association. Reproduced with permission.

4.1. Modified from figure 28 in Baerends, G. P. 1941. "Fortpflanzungsverhalten und Orientierung der Grabwespe *Ammophila campestris.*" *Tijdschrift voor Entomologie* 84: 71–248.

4.2. Image design by Vince Gallo, published as figure 4 in Gallo, V., Chittka, L. 2018. "Cognitive aspects of comb-building in the honeybee?" *Frontiers in Psychology* 9:900. DOI: 10.3389/fpsyg.2018.00900.

4.3. Photos by Florian Schiestl (bumble bee, white flower), Steve Johnson (sphingid moth), Scott Hodges (red flower), Rob Raguso (ipomoea); figure design modified from figure 1 in Clare, E. L., Schiestl, F. P., Leitch, A. R., Chittka, L. 2013. "The promise of genomics in the study of plant-pollinator interactions." *Genome Biology* 14: 207. DOI: 10.1186/gb-2013-14-6-207.

5.1. Photo by Jeremy Early, as figure 1A in Collett, M., Chittka, L., Collett, T. S. 2013. "Spatial memory in insect navigation." *Current Biology* 23 (17): R789–800. DOI: 10.1016/j.cub.2013.07.020

5.2 A and B. Figure design by Jürgen Tautz and Marco Kleinhenz (modified), published in Chittka, L. 2004. "Dances as windows into insect perception." *PLOS Biology* 2: 898–900. DOI: 10.1371/journal.pbio.0020216.

5.2. C. Redrawn after Barron, A. B., Plath, J. A. 2017. "The evolution of honey bee dance communication: a mechanistic perspective." *Journal of Experimental Biology* 220: 4339–46. DOI: 10.1242/jeb.142778.

6.1. Image redrawn after figure 1 from Collett, T., Kelber, A. 1988. "The retrieval of visuo-spatial memories by honeybees." *Journal of Comparative Physiology A* 163: 145–50. DOI: 10.1007/BF00612004.

6.2. Redrawn after figure 1 from Bennett, A.T.D. 1996. "Do animals have cognitive maps?" *Journal of Experimental Biology* 199: 219–24. DOI: 10.1242/jeb.199.1.219.

6.3. Photo by Lars Chittka, published as figure 1 in Skorupski, P., MaBouDi, H., Galpayage Dona, H. S., Chittka, L. 2018. "Counting insects." *Philosophical Transactions of the Royal Society B—Biological Sciences* 373: 20160513. DOI: 10.1098/rstb.2016.0513.

6.4. Design by HaDi MaBouDi, published as figure 5 in Skorupski, P., MaBouDi, H., Galpayage Dona, H. S., Chittka, L. 2018. "Counting Insects." *Philosophical Transactions of the Royal Society B—Biological Sciences* 373: 20160513. DOI: 10.1098/rstb.2016.0513.

6.5. A. Redrawn from figure 1 in Muller, M., Wehner, R. 1988. "Path integration in desert ants, *Cataglyphis fortis.*" *Proceedings of the National Academy of Sciences of the USA* 85: 5287–90. DOI: 10.1073/pnas.85.14.5287.

6.5. B–D. Redrawn after figure 2 from Chittka, L., Kunze, J., Shipman, C., Buchmann, S. L. 1995. "The significance of landmarks for path integration of homing honey bee foragers." *Naturwissenschaften* 82: 341–43. DOI: 10.1007/BF01131533.

6.6. Left: Photo by Joseph Woodgate, published in Woodgate, J. L., Makinson, J. C., Rossi, N., Lim, K. S., Reynolds, A. M., Rawlings, C. J., Chittka, L. 2021. "Harmonic radar tracking reveals that honeybee drones navigate between multiple aerial leks." *iScience* 24 (6): 102499. DOI: 10.1016/j.isci.2021.102499.

6.6. Right: Photo by Lars Chittka. Published as figure 2A in Chittka, L. 2017. "Bee cognition." *Current Biology* 27 (19): R1049–53. DOI: 10.1016/j.cub.2017.08.008.

6.7. Image series by Joseph Woodgate, published as figure 2B–D in Chittka, L. 2017. "Bee cognition." *Current Biology* 27 (19): R1049–53. DOI: 10.1016/j.cub.2017.08.008, and based on data from: Woodgate, J. L., Makinson, J. C., Lim, K. S., Reynolds, A. M., Chittka, L. 2016. "Life-long radar tracking of bumblebees." *PLOS ONE* 11 (8): e0160333. DOI: 10.1371/journal.pone.0160333.

7.1. Left: Figure design by Beau Lotto, published as figure 4A in Lotto, R. B., Chittka, L. 2005. "Seeing the light: illumination as a contextual cue to color choice behavior in bumblebees." *Proceedings of the National Academy of Sciences of the USA* 102: 3852–56. DOI: 10.1073/pnas.0500681102.

7.1. Right: Photo by Klaus Lunau.

7.2. Top: Photo by Lars Chittka.

7.2. Bottom: Data and figure 4 from Chittka, L., Thomson, J. D. 1997. "Sensori-motor learning and its relevance for task specialization in bumble bees." *Behavioral Ecology and Sociobiology* 41: 385–98. DOI: 10.1007/s002650050400.

7.3. Adapted from Theobald, J. 2014. "Insect neurobiology: how small brains perform complex tasks." *Current Biology* 24: R528–29. DOI: 10.1016/j.cub.2014.04.015, which is in turn based on Nityananda, V., Skorupski, P., Chittka, L. 2014. "Can bees see at a glance?" *Journal of Experimental Biology* 217: 1933–39. DOI: 10.1242/jeb.101394.

7.4. Photos and electron micrographs are by Beverley Glover, originally published as figure 1 in Whitney, H., Chittka, L. 2007. "Warm flowers, happy pollinators." *Biologist* 54: 154–59.

7.5. Image by Brigitte Bujok, Marco Kleinhenz, Jürgen Tautz, previously published in Dyer, A. G., Whitney, H. M., Arnold, S.E.J., Glover, B. J., Chittka, L. 2006. "Bees associate warmth with flower colour." *Nature* 442: 525. DOI: 10.1038/442525a.

7.6. Redrawn from figure 4 in Chittka, L., Niven, J. 2009. "Are bigger brains better?" *Current Biology* 19: R995–1008. DOI: 10.1016/j.cub.2009.08.023, which is in turn a redrawing of data from Giurfa, M., Zhang, S., Jenett, A., Menzel, R., Srinivasan, M. V. 2001. "The concepts of 'sameness' and 'difference' in an insect." *Nature* 410: 930–33. DOI: 10.1038/35073582.

7.7. Redrawn from figure 1 in Chittka, L., Jensen, K. 2011. "Animal cognition: concepts from apes to bees." *Current Biology* 21: R116–19. DOI: 10.1016/j.cub.2010.12.045, which shows the experimental procedure in: Avarguès-Weber, A., Dyer, A. G., Giurfa, M. 2011. "Conceptualization of above and below relationships by an insect." *Proceedings of the Royal Society of London B—Biological Sciences* 278 (1707): 898–905. DOI: 10.1098/rspb.2010.1891.

8.1. Redrawn after figure 1 from Leadbeater, E., Dawson, E. H. 2017. "A social insect perspective on the evolution of social learning mechanisms." *Proceedings of the National Academy of*

*Sciences of the USA* 114: 7838–45. DOI: 10.1073/pnas.1620744114, which is in turn a rendition of the experimental procedure used in: Dawson, E. H., Avarguès-Weber, A., Chittka, L., Leadbeater, E. 2013. "Learning by observation emerges from simple associations in an insect model." *Current Biology* 23: 727–30. DOI: 10.1016/j.cub.2013.03.035.

8.2. Photos by Ellouise Leadbeater.

8.3. Left: Photo series by Sylvain Alem; originally published as figure 3 in: Chittka, L. 2017. "Bee cognition." *Current Biology* 27 (19): R1049–53. DOI: 10.1016/j.cub.2017.08.008.

8.3. Right: figure 5A from: Alem, S., Perry, C. J., Zhu, X., Loukola, O. J., Ingraham, T., Søvik, E., Chittka, L. 2016. "Associative mechanisms allow for social learning and cultural transmission of string pulling in an insect." *PLOS Biology* 14 (10): e1002564. DOI: 10.1371/journal. pbio.1002564.

8.4. Photo is by Iida Loukola; other images redrawn from figure 2 in: Loukola, O. J., Solvi, C., Coscos, L., Chittka, L. 2017. "Bumblebees show cognitive flexibility by improving on an observed complex behavior." *Science* 355 (6327): 833–36. DOI: 10.1126/science.aag2360.

8.5. Top: Photo by Rotraut Sachs, published as figure 7A in Leadbeater, E., Chittka, L. 2007. "Social learning in insects—from miniature brains to consensus building." *Current Biology* 17: R703–13. DOI: 10.1016/j.cub.2007.06.012.

8.5. Bottom: Modified from figure 5 in Seeley, T., Buhrman, S. 1999. "Group decision making in swarms of honey bees." *Behavioral Ecology Sociobiology* 45: 19–31. DOI: 10.1007/s002650050536. Copyright © 1999 by Springer Nature. Reprinted with permission.

9.1. Top: Figures 2 and 5 from Dujardin, F. 1850. "Mémoire sur le systeme nerveux des insectes." *Annales des Sciences Naturelles B—Zoologie* 14: 195–206.

9.1. Bottom: Figure design published as figure 1D in Rother, L., Kraft, N., Smith, D. B., el Jundi, B., Gill, R. J., Pfeiffer, K. 2021. "A micro-CT-based standard brain atlas of the bumblebee." *Cell and Tissue Research* 386: 29–45. DOI: 10.1007/s00441-021-03482-z.

9.2. Original drawing from Kenyon, F. C. 1896. "The brain of the bee—a preliminary contribution to the morphology of the nervous system of the Arthropoda." *Journal of Comparative Neurology* 6: 134–210. DOI: 10.1002/cne.910060302.

9.3. Left: Original drawing from Ramón y Cajal, S., Sánchez, D. 1915. "Contribución al conocimiento de los centros nerviosos de los insectos." *Trabajos del Laboratorio de Investigaciones Biológicas de la Universidad de Madrid* 13: 1–68.

9.3. Right: Panels A–F of figure 4 from Paulk, A. C., Phillips-Portillo, J., Dacks, A. M., Fellous, J. M., Gronenberg, W. 2008. "The processing of color, motion and stimulus timing are anatomically segregated in the bumblebee brain." *Journal of Neuroscience* 28 (25): 6319–32. DOI: 10.1523/JNEUROSCI.1196-08.2008. (Copyright 2008 Society for Neuroscience.)

9.4. Modified from figure 1A in Hammer, M. 1993. "An identified neuron mediates the unconditioned stimulus in associative olfactory learning in honeybees." *Nature* 366: 59–63. DOI: 10.1038/366059a0. Copyright © 1993 by Springer Nature. Reprinted with permission.

9.5. Modified from Honkanen, A., Adden, A., Freitas, J. D., Heinze, S. 2019. "The insect central complex and the neural basis of navigational strategies." *Journal of Experimental Biology* 222 (Suppl. 1): jeb188854. DOI: 10.1242/jeb.188854.

10.1. Left, top and bottom: Photos by Lars Chittka.

10.1. Right: Modified from figure 5 in Stelzer, R. J., Stanewsky, R., Chittka, L. 2010. "Circadian foraging rhythms of bumblebees monitored by radio-frequency identification." *Journal of Biological Rhythms* 25: 257–67. DOI: 10.1177/0748730410371750.

10.2. Photo by Helga Heilmann, previously published as figure 1 in Chittka, A., Chittka, L. 2010. "Epigenetics of royalty." *PLOS Biology* 8: e1000532. DOI: 10.1371/journal.pbio.1000532.

10.3. Electron micrographs by Johannes Spaethe, published as figure 1A in Spaethe, J., Chittka, L. 2003. "Interindividual variation of eye optics and single object resolution in bumblebees." *Journal of Experimental Biology* 206: 3447–53. DOI: 10.1242/jeb.00570.

10.4. Top: Modified from figure 1 in Saleh, N., Chittka, L. 2007. "Traplining in bumblebees (*Bombus impatiens*): a foraging strategy's ontogeny and the importance of spatial reference memory in short range foraging." *Oecologia* 151: 719–30. DOI: 10.1007/s00442-006-0607-9.

10.4. Bottom: Image designs by Joseph Woodgate, previously published as figure 1 in Woodgate, J. L., Makinson, J. C., Lim, K. S., Reynolds, A. M., Chittka, L. 2016. "Life-long radar tracking of bumblebees." *PLOS ONE* 11 (8): e0160333. DOI: 10.1371/journal.pone.0160333.

10.5. Redrawn after data from Chittka, L., Dyer, A. G., Bock, F., Dornhaus, A. 2003. "Bees trade off foraging speed for accuracy." *Nature* 424: 388. DOI: 10.1038/424388a.

10.6. Based on data from Raine, N. E., Chittka, L. 2008. "The correlation of learning speed and natural foraging success in bumble-bees." *Proceedings of the Royal Society of London B—Biological Sciences* 275: 803–8. DOI: 10.1098/rspb.2007.1652.

10.7. Rearranged from figure 1 in Li, L., MaBouDi, H., Egertová, M., Elphick, M. R., Chittka, L., Perry, C. J. 2017. "A possible structural correlate of learning performance on a colour discrimination task in the brain of the bumblebee." *Proceedings of the Royal Society of London B—Biological Sciences* 284 (1864): 20171323. DOI: 10.1098/rspb.2017.1323.

11.1. Photo by Lars Chittka, previously published as figure 1 in Mendl, M., Paul, E. S., Chittka, L. 2011. "Animal behaviour: emotion in invertebrates?" *Current Biology* 21: R463–65. DOI: 10.1016/j.cub.2011.05.028.

11.2. Photos and images by Thomas Ings. Line art is from: Ings, T. C., Wang, M. Y., Chittka, L. 2012. "Colour independent shape recognition of cryptic predators by bumblebees." *Behavioral Ecology and Sociobiology* 66: 487–96. DOI: 10.1007/s00265-011-1295-y; results are based on data from: Ings, T. C., Chittka, L. 2008. "Speed-accuracy tradeoffs and false alarms in bee responses to cryptic predators." *Current Biology* 18: 1520–24. DOI: 10.1016/j.cub.2008.07.074.

11.3. Figure design by Cwyn Solvi, based on data in Solvi, C., Baciadonna, L., Chittka, L. 2016. "Unexpected rewards induce dopamine-dependent positive emotion–like state changes in bumblebees." *Science* 353 (6307): 1529–31. DOI: 10.1126/science.aaf4454.

11.4. Figure design by Joanna Brebner, previously published as figure 1 in Brebner, J. S., Chittka, L. 2021. "Animal cognition: the self-image of a bumblebee." *Current Biology* 31 (4): R207–9. DOI: 10.1016/j.cub.2020.12.027.

11.5. Photos by Lars Chittka.

12.1. Photo by Joseph Wilson.

# INDEX

Page numbers in *italics* refer to figures.